The World in Which We Occur

The World in Which We Occur

John Dewey, Pragmatist Ecology, and American Ecological Writing in the Twentieth Century

NEIL W. BROWNE

The University of Alabama Press
Tuscaloosa

Copyright © 2007
The University of Alabama Press
Tuscaloosa, Alabama 35487-0380
All rights reserved
Manufactured in the United States of America

Typeface: Baskerville

∞

The paper on which this book is printed meets the minimum requirements of American National Standard for Information Sciences–Permanence of Paper for Printed Library Materials, ANSI Z39.48-1984.

Library of Congress Cataloging-in-Publication Data

Browne, Neil W.
 The world in which we occur : John Dewey, pragmatist ecology, and American ecological writing in the twentieth century / Neil W. Browne.
   p. cm.
 Includes bibliographical references and index.
 ISBN-13: 978-0-8173-1581-8 (alk. paper)
 ISBN-10: 0-8173-1581-0 (alk. paper)
 1. Human ecology in literature. 2. Human ecology—Philosophy. 3. Dewey, John, 1859–1952—Criticism and interpretation. I. Title. II. Title: John Dewey, pragmatist ecology, and American ecological writing in the twentieth century.
 PN48.B66 2007
 810.9'36—dc22
                                                        2007007326

Excerpts from "Meditation on a Skull Carved in Crystal" and "Rain Country" copyright 1993 by John Haines. Reprinted from *The Owl in the Mask of the Dreamer* with the permission of Graywolf Press, Saint Paul, Minnesota. "The Tundra" by John Haines, from *Winter News* (Wesleyan University Press, 1966). Copyright 1966 by John Haines and reprinted by permission of Wesleyan University Press.

A portion of this book appeared in *ISLE: Interdisciplinary Studies in Literature and Environment* 11.2 under the title "Activating the 'Art of Knowing': John Dewey, Pragmatist Ecology, and Environmental Writing." I am grateful for permission to reprint that material at this time.

> If the gap between organism and environment is too wide, the creature dies.
> —John Dewey

Contents

List of Illustrations
ix

List of Abbreviations for Works of John Dewey
xi

Acknowledgments
xiii

Introduction: John Dewey and Pragmatist Ecology
1

1. An Arc of Discovery: John Muir's *My First Summer in the Sierra*
21

2. "The Form of the New": Pragmatist Ecology and *Sea of Cortez*
50

3. Rachel Carson's Marginal World:
Pragmatist Ecology, Aesthetics, and Ethics
78

4. "The Coldest Scholar on Earth": Silence and Work in John Haines's
The Stars, the Snow, the Fire
111

5. Northern Imagination: Wonder, Politics, and Pragmatist Ecology
in Barry Lopez's *Arctic Dreams*
143

Conclusion: (Eco)logic in the Utah Landscape
169

Notes
187

Works Cited
205

Index
219

Illustrations

John Dewey in Hubbards, Nova Scotia, mid-1940s
1

Hetch Hetchy Valley, 1911
20

John Steinbeck and Sparky Enea on the bridge of the *Western Flyer*, 1940
50

Rachel Carson at a microscope, 1951
78

Former Richardson Roadhouse
111

Iceberg
143

Wood ibis, scarlet ibis, flamingo, white ibis, by Alexander Wilson
169

Abbreviations for Works of John Dewey

AE	*Art as Experience* (1934), *Later Works*, vol. 10.
"Americanism"	"Americanism and Localism" (1920), *Middle Works*, 12: 12–16.
"Darwinism"	"The Influence of Darwinism on Philosophy" (1909), *Middle Works*, 4: 3–14.
DE	*Democracy and Education* (1916), *Middle Works*, vol. 9.
"Democracy"	"Philosophy and Democracy" (1919), *Middle Works*, 11: 41–53.
EN	*Experience and Nature* (1925), *Later Works*, vol. 1.
"Freedom"	"Freedom" (1937), *Later Works*, 11: 247–55.
"Future"	"Philosophy's Future in Our Scientific Age" (1949), *Later Works*, 16: 369–82.
KK	*Knowing and the Known* (1949), *Later Works*, 16: 1–294.
Logic	*Logic: The Theory of Inquiry* (1938), *Later Works*, vol. 12.
"Need"	"The Need for a Recovery of Philosophy" (1917), *Middle Works*, 10: 3–48.
PP	*The Public and Its Problems* (1927), *Later Works*, 2: 235–372.
"Pragmatism"	"The Development of American Pragmatism" (1925), *Later Works*, 2: 3–21.
QC	*The Quest for Certainty* (1929), *Later Works*, vol. 4.
"Task"	"Creative Democracy—The Task Before Us" (1939), *Later Works*, 14: 224–30.

Acknowledgments

In 1990, when I returned to university study after a period of about twelve years, two professors were instrumental in deeply changing my intellectual life: John Schell and Bernd Decker. Their influence abides. I would also like to thank those who, whether they know it or not, have contributed to my thinking, writing, and academic life: Lothar Hönnighausen, Claus Daufenbach, David Heaton, Ken Daley, Josie Bloomfield, Matthew Cooperman, Christine Gerhardt, Kathleen Dean Moore, Glen Love, and Scott Slovic. Louise Westling, in what seems a long while ago, suggested the environmental possibilities in John Dewey's *Art as Experience*. I have been reading Dewey ever since. My colleagues at Oregon State University–Cascades and in the Oregon State University English Department have been and remain a pleasure to work with, and I have greatly benefited from their commitment to scholarly and creative work. A special thanks to Natalie Dollar, James Foster, Sandy Brooke, Kerry Ahearn, and David Robinson. I owe an unusually large debt to Henry Sayre for his expert reading of my work, for his knowledge of American art, for his solid friendship, and for his good cooking. The two readers for the University of Alabama Press made excellent suggestions toward revision of the manuscript, and it is a far better book for them. All errors are of course my own. The people at the University of Alabama Press have been a pleasure to work with. Finally, Robert DeMott has been my mentor, teacher, and dear friend for many years now. He will always be a force in my life. Many thanks.

I owe immeasurable thanks to my mother, Joan Browne, who has helped me along, sometimes against all odds, my entire life. She has never lost faith. I kept my brother, Peter, firmly in mind when I was thinking about my audience. My daughter, Sarah Cumbie, will always be an inspiration. I am especially thankful to my wife, Terri Cumbie, to whom I dedicate this book, for her deep love and companionship. I thank her too for her steadfast social and environmental conscience. She is also a very able copy editor. She never fails to make my work and my life whole.

The World in Which We Occur

John Dewey in Hubbards, Nova Scotia, mid-1940s. Courtesy of Special Collections Research Center, Morris Library, Southern Illinois University, Carbondale.

Introduction
John Dewey and Pragmatist Ecology

Everything that exists in as far as it is known and knowable is in interaction with other things.

—John Dewey

The career and destiny of a living being are bound up with its interchanges with its environment, not externally but in the most intimate way.

—John Dewey

John Dewey's insistence that human experience is inextricable from the nonhuman world—from the world of other things and environments—provides a key to how his thought can help us imagine the present and future role of human culture in the world's ecologies. The other-than-human world participates in all human experience, and Dewey's philosophy can aid in our articulation of ways of being that honor the contribution of the nonhuman world to our everyday experience. For Dewey, the acme of human existence—our being—is implicated in aesthetic experience, and Deweyan aesthetic experience is, radically, most often rooted in everyday life, itself rooted in everyday environments. Aesthetic can be understood as ecological. Our daily lives, communities, ecologies, even our simplest acts, are beautiful on a level with the most treasured works of art. Moral and ethical acts can be aesthetic; art can be moral.[1] Given Dewey's emphasis on environment and community, to consummate human experience at its highest potential demands both an ecologically intact world and a renewed democratic culture. Both requirements are related in that they imply a deep respect for the human and nonhuman other, and such respectfulness insists that we engage our imaginations with the world to our utmost ability.[2] In Dewey's pragmatism, imagination is "the capacity to understand the actual in light of the possible" (Alexander, "Moral Imagination" 371), and to wield imagination most effectively in our young century requires access to vast amounts of information from disparate sources, sources very often cordoned off from the public. I am calling this informed process of understanding and imagining "the actual" in terms of "the possible" pragmatist ecology. Pragmatist ecology understands that, in relation to physical environments in which humans are involved, crucial roles are played not only by the biology of creatures but also by the culture of the human creature. By "involved" I mean more than a physical impact on an environment. Writing, visual art, and music as well as animal populations, drainage regimes, and the hydrological cycle are crucial to understanding our roles in any environment. As Mark Allister argues in his informative study on nature writing and autobiography, *Refiguring the Map of Sorrow*, "most theorizing about nature writing, because of its preoccupation with 'environmental issues' or science and its desire to keep humans completely away as a defining characteristic, needs altering as well as enlarging" (33). Pragmatist ecology is a way to extend our thinking about the interrelationships between human culture and the physical world. Pragmatist ecology looks at the relations among art, science, politics, intelligence, and the physical world, rendering the artificial boundaries separating them porous—ecotonal. An ecotone is a transitional zone between ecosystems, such as the tidal zones or the edges between a field and forest. These are places of intensified energy, where genetic exchange and evolutionary potential are initiated. This heightened potential exists also in cultural ecotones.[3]

Effectively rethinking and reimagining what a culture underpinned by environmentally sound and respectful values might look like in the United States in particular will draw from within U.S. culture itself, and the investigation of public intelligence has been an important task of the American philosophical tradition of pragmatism, especially in Dewey's work. This idea of public intelligence becomes not only important to the understanding of what an environmentally principled culture might look like, but also essential to the continuity of our imperiled democratic system of government. So in this sense, a substantial link can be drawn between ecology and democracy, a claim that runs through most everything I write in this book. At his core, Dewey was deeply concerned with democracy and with creating a public—and in turn a public intelligence—prepared to participate meaningfully in a democratic culture. He often calls this public intelligence the "art of knowing," and the art of knowing depends on interrelationships between individuals and various ways of meaning. In other words, if knowledge and individuals are too specialized and isolated, attempts at disseminating knowledge among the larger public will fail. Extreme individualism and the accumulation of knowledge by elite groups impede the attempt to create a public intelligence. In *Democracy and Education* (1916), Dewey writes: "From a social standpoint, dependence denotes a power rather than a weakness; it involves interdependence. There is always a danger that increased personal independence will decrease the social capacity of an individual. In making him more self-reliant, it may make him more self-sufficient; it may lead to aloofness and indifference. It often makes an individual so insensitive to his relations to others as to develop an illusion of being really able to stand and act alone—an unnamed form of insanity which is responsible for a large part of the remedial suffering of the world" (49). Dewey questions the cherished ideal of self-sufficiency and finds this most stubborn of U.S. cultural assumptions detrimental to society itself. Blindness to interrelationships and relations produces a form of insanity leading to widespread misery. This ideal of self-sufficiency on a national level, coupled with over consumption of natural resources, has contributed to preemptive war, the isolation of the United States from the rest of the world, and the accelerated assault on the environment by energy-industry-controlled government. This is, indeed, a form of insanity in our own time—a remedial one, one hopes.

Granted, in 1916 Dewey was talking about "social intercourse" (*DE* 48), but later in his career he extended these ideas to ways of thinking that embrace both cultural and physical environments, to "the art of knowing." To divvy up the arts of knowing into disciplines that are unable or unwilling to speak to one another or to the public, that cannot recognize their interdependence and their relationship to the larger culture, fosters a host of miseries. In 1949, a few years before his death at the age of ninety-two, Dewey remained consistent in his belief that cultural troubles are remedial and that their remedy depends on a kind of public intelligence. He writes in "Philosophy's Future in Our Scientific Age": "But it is surprising that those who call themselves 'liberals' should fail to see that the *absence of a knowledge genuinely humane* is a great source of our remedial troubles, and that its active presence is needed in order to translate the articles of their faith into works" ("Future" 375, Dewey's emphasis). Addressing these troubles throughout his career, Dewey always insisted that knowledge be put to use; the primary use of that knowledge now should be to create a humane democratic culture grounded in ecological principles.

To accomplish this goal, we must first begin to break down rigid boundaries and replace them with permeable ones. Dewey exposes as human constructions (mostly self-serving) the impermeable dichotomies and hierarchies between science and art that many still seem to want to shore up.[4] In *Experience and Nature,* Dewey uncovers and contests this dichotomy:

> The failure to recognize that knowledge is a product of art accounts for an otherwise inexplicable fact: that science lies today like an incubus upon such a wide area of beliefs and aspirations. To remove the deadweight, however, recognition that it is an art will have to be more than a theoretical avowal that science is made by man for man, although such recognition is probably an initial preliminary step. But the real source of the difficulty is that the art of knowing is limited to such a narrow area. Like everything precious and scarce,

> it has been artificially protected; and through this very protection it has been dehumanized and appropriated by a class. As costly jewels of jade and pearl belong only to a few, so with the jewels of science. The philosophic theories which have set science on an altar in a temple remote from the arts of life, to be approached only with particular rites, are a part of the technique of retaining a secluded monopoly of belief and intellectual authority. (286)

First of all, the separation of art from science is fallacious from the outset. Science is not a value-neutral, objective way of knowing the world, but a tool constructed by human beings in order to communicate with one another. This is not to deny that science can be a primary way of knowing the world, but certainly it is not the only one, or even the best one. Art is also a product of human beings, and "thinking is pre-eminently an art; knowledge and propositions which are the products of thinking, are works of art, as much so as statuary and symphonies" (*EN* 283). Science and art and music are all aspects of the larger art of knowing, but here we begin to run into a problem that has not abated since Dewey's time, one that has proliferated to "an unnamed form of insanity" that causes avoidable suffering around the globe. This remedial suffering, of course, is the result of individuals and disciplines believing they are independent of one another.

A vibrant, participatory democracy cannot abide entities in isolation any more than ecosystems can, and at the core of Dewey's thought lies the belief in democracy as a way of life, not just a political form.[5] Dewey writes in *The Public and Its Problems:*

> Singular things act, but they act together. Nothing has been discovered which acts in entire isolation. The action of everything is along with the action of other things. The "along with" is of such a kind that the behavior of each is modified by its connection with others. There are trees which can grow only in a forest. Seeds of many plants can successfully germinate and develop only under conditions furnished by the presence of other plants. Reproduction of kind is dependent upon the activities of insects which bring about fertilization. The life history of an animal cell is conditioned upon connection with what other cells are doing. Electrons, atoms and molecules exemplify the omnipresence of conjoint behavior. (250)

Dewey's notion of "along with" is vital. Any art of knowing depends upon interactions among creatures. All knowing is a kind of sharing, of being "along with" another. This "along with" leads to vibrant community members able to understand that their actions are of necessity integral to the well-being of their neighbors, or to their detriment. Dewey supports his argument for responsible participation within a democratic community with examples from nature, but in an even more important sense, examples are drawn from an understanding of nature informed by science, right down to the near view of atoms and electrons. Terry Tempest Williams takes a strikingly similar stance, also making a direct link between the physical world and radical democracy: "The power of nature is the power of a life in association. Nothing stands alone. On my haunches, I see a sunburst lichen attached to limestone; algae and fungi are working together to break down rock into soil. I cannot help but recognize a radical form of democracy at play. Each organism is rooted in its own biological niche, drawing its power from its relationship to other organisms. An equality of being contributes to an ecological state of health and succession" (*Open Space* 58).[6] Williams's scientific background enables her to perceive in even the lichen, a hybrid form, a suggestion of the democratic process; democracy and ecology participate in creating a healthy condition. Integral to a democratic way of life is community access to the art of knowing, which demands, among other things, the availability of scientific knowledge to the larger community. It is necessary to remark here that when Dewey speaks of science, he means science as a method—as the interaction of creative intelligence with the physical world in order to better know that world—not the facts and findings of science. In *The Quest for Certainty* he argues that any kind of knowing is a process that results in action: "If we see that knowing is not the act of an outside spectator but of a participator inside the natural and social scene, then the true object of knowledge resides in the consequences of directed action" (157). Knowing is not a spectator sport; creative intelligence participates in the natural and social worlds into which it inquires. There is no disinterested, objective observer. The "true object of knowledge" is not a scientific fact but rather the consequence of "directed action": not facts isolated from one another and their larger contexts, but facts put into cultural use form knowledge. And in order for action to be "directed," the agent must have access to the art of knowing. In a Deweyan democratic culture, neither science nor art nor philosophy can be relegated to the realm of experts.

Dewey's work can help us understand the ecological importance of the accessibility of knowledge, and all the writers I discuss in this book—John Muir, John Steinbeck, Edward Ricketts, Rachel Carson, John Haines, Barry Lopez, and Terry Tempest Williams—work across or through boundaries between different ways of knowing in one way or another. I concentrate on nonfiction that strongly engages a particular environment, but I resist calling this work "nature writing" because that term tends to elide the crucial role of the human being participating in its environment. In *Conserving Words,* his rhetorical study of how writers have helped shape the environmental movement, Daniel Philippon notes: "The constituent parts of the term 'nature writing' address the dual issues with which the literary conversation it describes is fundamentally concerned: the definition of 'nature' and the problem of language. Nature writing might best be defined, in other words, in terms of its expansive subject: the interaction of nature and culture in a particular place" (10). While this is a workable definition, especially in its terms of interaction, it also leans toward dualism and a fundamental concern. To my mind, nothing is more important in this kind of writing than its practice of interaction and interrelation, and I try to keep my focus closely on this point. Therefore, I refer to this work as "ecological writing" because it understands the human world and the natural world as participating along with one another. The participatory valence of these texts is of utmost importance. My focus is not on the specifics of any one theory or way of knowing the world, but on a pragmatist understanding of the responsibility of literature and other means of inquiry—the art of knowing—to a democratic culture rooted in the physical world—on a pragmatist ecology, an ecology of progressive meaning. In this construction, ecology implies not only the scientific discipline but also its responsibility to a democratic culture at large.

In a recent study of ecocriticism and nature writing, *The Truth of Ecology,* Dana Phillips warns against the idea that literary criticism, literature, and ecology can inform one another. Phillips's book is a scorched-earth indictment of a certain brand of ecocriticism that harkens after a renewed realism and yearns for a lost pastoral experience. Much of what Phillips has to say in this study is well taken and aimed at rejuvenating ecocriticism—if not nature writing—for which he sees only a dim future. Phillips basically denies even the possibility that writing can be ecological because the science of ecology does not support such a view, and this slippage into dualism—in this case between art and science—is precisely what Dewey argued strenuously against throughout his career. Dualism diminishes both art and science. Nevertheless, Phillips embraces Dewey and neopragmatism for their sense of contingency and their understanding of truth as ongoing inquiry, but again, he seems to miss a key point in Dewey's thinking: the inextricable linkage of the cultural and the natural. For instance, Phillips claims that in the context of evolution, "The impact of the giant asteroid that may have caused the dinosaur's extinction and the buildup of greenhouse gasses that may cause yours are both cases of business as usual" (149). This ignores the fact that the problems caused by a giant asteroid crashing to earth and by greenhouse gases are fundamentally different. In the case of the latter, ethical and moral concerns are as much in evolutionary play as physical ones, maybe more so. Phillips goes on to argue that "ecocritics need to recognize that cultural and natural processes are functionally distinct or at least distant from one another, and that maintaining that distinction, and keeping the distance, is probably a good idea" (149).[7] But at least in the case of global warming, cultural and natural processes are inextricable. Dewey would disagree with Phillips: he would see that distance as dangerous to a vibrant culture. Dewey provides a firm philosophical ground for integrating science, ethics, the physical world, and art for a pragmatist ecology. At stake here is a resuscitation of democratic culture and a solid claim for the centrality of environmental values within that culture.

Phillips also takes ecocriticism to task for using ecology as a metaphor. This line of thinking is also a symptom of our cultural imperative to keep ways of thinking and knowing in separate spheres. Surely, much science is dependent on metaphor and employs it as an explanatory tool—the concept of "stream capture" comes immediately to mind. However, Phillips argues that "It is more productive, and more properly historical, to understand the development of ecology as a struggle to divest itself of analogical, metaphorical, and mythological thinking, and of literary means of suasion (including narrative)" (58). Besides betraying an urge to keep these diverse ways of knowing in separate boxes, Phillips's claim is at best only partly true. It is one part of the story. Ecologist T. F. H. Allen writes in "Community Ecology" that "Often the ultimate contribution of mathematical formalities is a justification for metaphor" (335). Along these same lines, Gary Snyder endorses the metaphorical valence of ecology: "Also, the term 'ecology,' which includes energy-exchange and interconnection, can be metaphorically extended to other realms. We speak of 'the ecology of the imagination' or even of language, with justification; 'ecology' is a short-hand term for complexity in motion" ("Ecology" 9).

Snyder's claim resonates with Dewey's argument in *Democracy and Education* that the remedial suffering of the world could be relieved through the acknowledgment of our interdependence upon one another and a complex world. To set science aloof from metaphor and narrative is to court disaster. If "ecology," with all its connotations of interrelationships and complexities, has conceptual links to works of the imagination and to the idea of democracy, why refrain from using those links to the utmost? As Philippon argues, "If we posit social transformation as a kind of 'social disturbance' (much like fire is a natural disturbance), we might see metaphor as the agent of that disturbance in the complex system I am calling the ecology of influence" (5). Metaphor in this sense participates in a kind of ecology that creates social change. With democratic, environmentally respectful cultures at stake, why, as Phillips seems to advocate, contract and compartmentalize thinking by banishing metaphor when metaphor is one of our key tools to expand our range of thought? To do so seems to support Dewey's claim that knowledge has been artificially protected and appropriated by a particular elite. Good Deweyan practice aims to break down these artificial barriers.

Dewey's philosophy is peppered with objections to the detrimental, antidemocratic consequences of overspecialization that contribute to the general degradation of meaningful experience in American culture. For example, John Steinbeck, Edward Ricketts, and Rachel Carson work in the littoral zones and tide pools, and in their work we find an interesting crossing between subject matter as transitional zone—an ecotone—and the writing itself as a mediational space between the humanities, science, and the general culture. For Dewey, available knowledge is a central need for a participatory democracy, so this notion of writing as a transitional space engages practical politics as well. Carson is a perfect example: her integration of literary language with solid science and ecological knowledge, along with a deep concern for the everyday person, leads to her powerful advocacy. Science, aesthetics, ethics, and advocacy are linked. For any kind of ecological awareness to evolve into informed advocacy or political efficacy in a broad-based democracy, citizens must have access to knowledge that has been appropriated by relatively few experts.

This knowledge is essential for meaningful experience. Experience is a key (and difficult) concept in Dewey's thought, so I will diverge here for a moment into experience. Experience is not something we have and look back on; rather, it is an ongoing process rooted in the physical world. In *Experience and Nature,* Dewey cites Charles Lyell's revitalization of geology: "Lyell revolutionized geology by perceiving that the sort of thing that can be experienced now in the operations of fire, water, pressure, is the sort of thing by which the earth took on its present structural forms" (12). The geological past participates in present experience. It is important to note here too that revolutionary scientific theory rests on Lyell's act of "perceiving," and for Dewey, similar to his definition of imagination, "to *perceive* is to acknowledge unattained possibilities; it is to refer the present to consequences, apparition to issue, and thereby to behave in deference to the *connections* of events" (*EN* 143, Dewey's emphasis). To see the actual in the light of the possible depends upon the perception of relations; hence experience "is *of* as well as *in* nature. It is not experience which is experienced, but nature—stones, plants, animals, diseases, health, temperature, electricity, and so on. Things interacting in certain ways *are* experience; they are what is experienced. Linked in certain other ways with another natural object—the human organism—they are *how* things are experienced as well. Experience thus reaches down into nature; it has depth. It also has breadth and to an indefinitely elastic extent. It stretches. That stretch constitutes inference" (*EN* 12–13, Dewey's emphasis). Experience *is* interaction and interrelatedness, and, clearly, the art of knowing, which necessarily includes inference, is an interaction always embedded in the physical environment. For Dewey, an integral element in the process of rendering experience meaningful is the perception of relations, especially among the organism and its environment. Experience is ongoing and natural, like drawing a breath, though conscious effort is required to make it meaningful:

> Experience like breathing is a rhythm of intakings and outgivings. Their succession is punctuated and made a rhythm by the existence of intervals, periods in which one phase is ceasing and the other is inchoate and preparing. William James aptly compared the course of a conscious experience to the alternate flights and perchings of a bird. The flights and perchings are intimately connected with one another; they are not so many unrelated lightings succeeded by a number of equally unrelated hoppings. Each resting place in experience is an undergoing in which is absorbed and taken home the consequences of prior doing, and, unless the doing is that of utter caprice or sheer routine, each doing carries in itself meaning that has been extracted and conserved. As with the advance of an army, all gains from what has been already effected are periodically

consolidated, and always with a view to what is to be done next. If we move too rapidly, we get away from the base of supplies—of accrued meanings—and the experience is flustered, thin, and confused. If we dawdle too long after having extracted a net value, experience perishes of inanition. (*AE* 62)

Again, experience requires above all conscious effort, and regardless of Dewey's unfortunate military metaphor, experience relies on an accrual and conservation of meanings, at least partly derived from knowledge that permeates the natural world and the cultural environment. Even one of experience's most basic components, perceiving, depends upon making connections, on turning apparitions into issues. Science is a primary supplier of the knowledge essential to experience, and for that knowledge to remain in the hands of a few frustrates the ability of the many to achieve meaningful experience, which then in turn impedes the ongoing contribution of knowledge from a wide spectrum of participants to the larger culture. Dewey warns that getting too far away from the base of supply—accrued meanings—inevitably results in diminished experience; surely, if the supply is pinched off, the same occurs. One vital difference, however, is that if we move too far ahead or remain inane, that is our fault. If supply is impeded by the intentional relegation of information to an elite few, that is a conscious impediment to the culture's ability to accrue meaning. No democratic culture can function vitally without access to the ways and rhythms of knowing.

Elaborating and examining the complex relationships among the humanities, science, and other ways of knowing can help build a culture rooted in ecological principles. In *Art as Experience,* Dewey's aesthetics points up the human being's close linkage and interpenetration with the nonhuman, and "Oppositions of mind and body, soul and matter, spirit and flesh all have their origin, fundamentally, in fear of what life may bring forth. They are marks of contraction and withdrawal" (28). On the other hand, pragmatist ecology imagines an outward step beyond our own parameters into the dynamic complexities of ecosystemic experience, a movement that calls for innovative literary and scientific forms that can articulate the interrelationships among literature, art, and science. In an essay on ethics and literary study, Derek Attridge argues: "innovative mental acts produce lasting alterations in the subjectivity that achieves them: once I have articulated the new thoughts that I had dimly apprehended, my thinking will never be entirely the same again. If that new articulation becomes public, with the disarticulation of settled modes of thought that made it possible (and thus that it made possible), it may alter cognitive frameworks across a wider domain, allowing for further acts of creativity in other minds" (22–23). Attridge's understanding of the process of changing subjectivities and cognitive frameworks complements Dewey's articulation of having an experience with its intakings and outgivings, with James's flights and perchings. Pragmatist ecology offers an articulation of ecological perception along with the aesthetic and ethical structures that can grow from that perception, and such an offering can, while availing itself of cultural forms, contribute to the disarticulation of settled modes that ignore ecological destruction, to alterations in subjectivity so that the human subject can better realize his or her role as an individual within natural communities, and to alterations of cognitive frameworks across a wider domain that can function as an awakening to culture's interrelationships with the physical environment. As Dewey suggests at the beginning of *Art and Experience,* ecology has a large part to play in the reawakening: "It is quite possible to enjoy flowers in their colored form and delicate fragrance without knowing anything about plants theoretically. But if one sets out to *understand* the flowering of plants, he is committed to finding out something about the interactions of soil, air, water, and sunlight that condition the growth of plants" (10, Dewey's emphasis). In other words, knowledge of the plant's ecology—perception of the relations among plant, soil, color, and human—contributes to meaningful experience.

Dewey moves beyond privileging experience as a purely subjective event; experience is "a product . . . of continuous and cumulative interaction of an organic self with the world" (*AE* 224). In opposition to a spectator theory of knowledge, he insists on experience as an interrelational process in which self and world engage. Central to his argument is the attempt to recover "the continuity of esthetic experience with normal processes of living" (16). In Dewey's aesthetics, writes Thomas M. Alexander, "The material out of which human life is built is 'experience,' understood in its Deweyan sense as that vast concurrence of natural events and cultural meanings in all their obscurity and power as well as in their focal clarity and luminosity. The tremendous task to be undertaken is to grasp the present—not as an immediate, isolated bare occurrence, as an indefinitely fleeting 'now,' but as the dynamically insistent occasion for establishing continuity or growth of meaning. Present experience stands for that whole complexity which establishes the human project as such" (*Horizon* 269). Experience is a confluence of relationships like the relationships among ecosystems we are just beginning to understand; additionally, ecosystems and experience undergo dynamic change over time. This is not to posit either experience or ecosystems as foundational. In recent ecological thought, the concept of patch dynamics has become a powerful way of envisioning not foundational but dynamic living systems. Patch dynamics insists upon smaller units of populations at different successional stages, such as an opening in a forest or shifting tide pools in the littoral zone. These patches together form a dynamic landscape, and the total of all disturbance tends to lead to a general, relative stability, or metastability.[8] As Dewey, too, well knew. "Any achieved equilibrium of adjustment with the environment is precarious because we cannot evenly keep pace with changes in the environment" ("Need" 9). The slippery relationship of writer, text, reader, and critic to a changing world is always a contingent one. Experience conceived of as a meaningful interrelationship with the world in all its cumulativeness suggests that each moment of human interface with the environment contains the potentiality of aesthetic expression of the experience, which, following Dewey's emphasis on the "normal processes of living," suggests an ecosystemic relationship among the creative human agent, cumulative human experience, art, and the natural world. Dewey's valorization of aesthetic experience has far-reaching ramifications for a heightened appreciation of the meaning of ecological writing to contemporary American culture, culture not "understood as schematic systems of oppositions but as shared significant responses to the world" (Alexander, *Horizon* 270). Again, "Experience occurs continuously, because the interaction of live creature and environing conditions is involved in the very process of living" (*AE* 42), and this very process of living has potential aesthetic significance. Ecological writing at its best enacts this dynamic sharing of significance, and for Dewey, "Shared experience is the greatest of human goods" (*EN* 157).

Dewey's inclusive interpretation of experience underwrites my contention that in ecological writing we witness the attempt to articulate the complex interaction and interpenetration of human culture and the physical world. I want to reiterate here that "experience is a matter of the interaction of an organism with its environment, an environment that is human as well as physical, that includes the materials of tradition, and institutions as well as local surroundings" (*AE* 251). The physical environment is not merely the setting in which experience begets experience, comes into being: it participates in the interaction itself, the interpenetration, the relationship between human culture and the physical world.

For example, Walden Pond, where Thoreau lived for two years, the book *Walden*, published in 1854, and twenty-first-century libraries all over the world that shelve the book participate in a certain ecology. For a closer look at this relationship, take Thoreau's classic description of the melting sandbank at the end of the "Spring" in *Walden*.[9] First of all, Thoreau acknowledges that the effect and frequency of the phenomenon he observes are heightened by human manipulation of the land: "the number of exposed banks of the right material must have been greatly multiplied since railroads were invented" (203). Both human culture and the sandy soil profile of the land surrounding Concord enhance the spectacle of the sandbank, but the most productive ecotone in this passage is not the boundary between the human railroad and the environment through which it cuts, but the one that exists between Thoreau's imagination and the material of the sandbank.

Thoreau gives us the composition of flowing materials that "overlap and interlace"—a mixture of sand and clay—and names it a "hybrid product," a gathering together of separate things. Since it is part mineral and part vegetable, it responds to a mutual influence, obeying "half way the law of currents and half way that of vegetation" (203). Thoreau introduces here a more complex conceptual ecotone. We have little trouble perceiving an ecotone between a forest and a field or a forest and a railroad cut. Thoreau, however, takes the idea a step further. He insists upon an interpenetration of mineral and vegetable that begins to defeat our received categories. He calls our basic morphologies into question. He brings his imagination to bear on the flowing sand and clay, and he is reminded of lichens and "of coral, of leopard's paws or bird's feet, of brains or lungs or bowels, and excrements of all kinds" (203). The movement shifts from simple life-forms to more complex animals, and then, surprisingly, his focus narrows to the internal organs of beasts, then to excrement in this strange meditation. It is a shift from external to internal and then to the internal made external. We forget we are looking at a sandbank. All manner of borders are rendered permeable—animal, vegetable, mineral, internal, and external—and the flowing sand becomes a conglomerate of life, down to the viscera that sustain life, and life's more disagreeable processes. Thoreau stands awed and tells us: "I feel as if I were nearer to the vitals of the globe, for this sandy overflow is something such a foliaceous mass as the vitals of the animal body. You find thus in the very sands an anticipation of the vegetable leaf. No wonder that the earth expresses itself outwardly in leaves, it so labors with the idea inwardly" (204).[10]

He also finds it expressive of language. The inside of a book, of course, also consists of leaves, and Thoreau literally begins to read the earth, connecting mineral, vegetable, and animal worlds with a playful and idiosyncratic sound and typographical symbolism:

> The overhanging leaf sees here its prototype. *Internally*, whether in the globe or animal body, it is a moist thick *lobe*, a word especially applicable to the liver and lungs and the *leaves* of fat (λειβω, *labor, lapsus*, to flow or slip downward, a lapsing; λοβος, *globus*, lobe, globe; also lap, flap, and many other words), *externally*, a dry thin *leaf*, even as the *f* and *v* are a pressed and dried *b*. The radicals of lobe are *lb*, the soft mass of the *b* (single lobed, or B, double lobed) with the liquid *l* behind it pressing it forward. (204)

Thoreau is obviously delighted by the sounds of human language. Language becomes the primary focus, and the liquid *l* pushing the lobe is an obvious reenactment of the flowing sand and clay. He sees into the internal workings of the globe and finds the alphabet. The very basic structure of Western language and expression—the alphabet—finds itself embedded in the bowels of the earth in a powerful expression of how human culture, writing, and world interpenetrate.

This whole passage may be seen as a trope for ecological writing in general. It is remarkable because it posits the earth in clearly textual terms, yet those terms are to a large degree indecipherable, a mixture of alphabets and hieroglyphics awaiting some "Champollion." The text, in order to emerge from the sandbank, demands the geomorphology of the railroad cut, a witnessing subject, and interpreters. Thoreau as witnessing subject in this passage submerges the narrating self as it approaches the "vitals of the globe" and implicitly calls upon the reader to perform Champollion's labor. In this sense, Thoreau's primary experience in nature becomes a virtual experience contained in a symbolic text, and once the text is read, the experience contributes to the ongoing aesthetic experience of the reader. What happens here is similar to Dewey's unusual definition of art: "Art is a quality that permeates an experience; it is not, save by a figure of speech, the experience itself. Esthetic experience is always more than esthetic" (*AE* 329).[11] To echo Attridge, innovative ways of thinking—creative expression in this case—can cause changes in subjectivities across wide cultural domains. Just as the flowing sand became language and thus permeated Thoreau's experience, the art of *Walden* permeates the reader's experience, functioning in this way as a mediational site between the physical world, the text, and experience. It is a giving and receiving across time and space, and *Walden* is one of the rare texts that seems in this way to constitute an ecology wherever and whenever the book is taken off the shelf. Books, then, like "immediate empirical things are just what they always were: endings of natural histories" (*EN* 110). I would add that they are also the beginnings of natural histories. Art is an active principle. So, to focus Dewey and Attridge a bit more closely on the concerns of this book, art with a decidedly ecological valence can theoretically permeate experience with a renewed consciousness of the ecological role of human culture in nature.

But, however convincing we may find Thoreau, we cannot will away the fact that the construction of the "ways in which we believe and expect" and the ways in which we perceive the natural world are social (*EN* 23). Dewey knows that "we believe many things not because the things are so, but because we have become habituated through the weight of authority, by imitation, prestige, instruction, the unconscious effect of language, etc." (23). In response to this aspect of social construction, he resists philosophical and critical methods that are "employed to cast doubt upon the reality of things external to mind and to selves, things and properties that are perhaps the most salient characteristics of ordinary experience" (38–39). Again, the physical world wields substantial power in a Deweyan construal of ordinary experience, and in order to enhance ordinary experience the integrity of both the world and the perceiving self must be sustained. This is not to deny social construction. As Lawrence Buell so cogently argues in *The Environmental Imagination,* "From an ecocentric standpoint a criterion built on a theoretical distinction between human constructedness and nonhuman reality . . . is far more productive than a criterion based on the presupposition of the inevitable dominance of constructedness alone" (113–14).[12] The critical privileging of this dominance mistakes "the structure of discourse for the structure of things" (*EN* 135). Nonhuman reality, the structure of things, must reenter the discussion of literature and culture. Dewey insists that "refined methods and products" (such as literary and cultural theory and criticism) "be traced back to their origin in primary experience, in all its heterogeneity and fullness; so that the needs and problems out of which they arise and which they have to satisfy to be acknowledged" (39). Primary experience is lodged in the relation between the human creature and the environment. Both the social and the natural participate in experience.

David Mazel, a critic committed to environmental issues—if not environmentalism as it is mainly practiced in the United States—argues that any ecocritical endeavor "that treats nature itself as a preexisting given rather than a discursively constructed artifact" (28) is misguided; he insists that Yosemite, for instance, be considered a "continually reconstituted social text" (159). While I am sympathetic to Mazel's attempt to uncover past horrors underlying the rhetoric of wilderness and the practice of preservation, I am not ready to concede him the physical world entirely. Anna L. Peterson articulates the risk here to an ethical relation to the physical world: "However, extreme constructionist positions abstract the production of meaning from the natural and material conditions of human life. Humanness is defined not by the dialectical interaction between persons and the world but by the largely interior process of self-creation through culture and language" (59). If our focus is inward, our conception of ethics comes from only the self, but if our focus is on the interaction between the human and the nonhumans, as Dewey would certainly approve, our ethical stance is predicated on our perception of relations between the human and the rest of the world. As Larry A. Hickman argues, "Nature is a construct, or cultural artifact, but it has not been constructed out of nothing" (53). So, certainly, we often turn the world into a text. The world as a text or palimpsest, for instance, is one of John Muir's favorite tropes, and as we have seen, Thoreau often thought of the world in textual terms. But Muir never felt he could successfully read that text, and the world often remained an undeciphered hieroglyphic for Thoreau. In Mazel's book, the critic is better able to read the world than either Thoreau or Muir, and he does so in an interesting, productive way, but that reading is costly. In his reading, the physical world becomes a correlate of human discourse and practice, and this placement of human discourse exclusively at the center of things threatens to objectify the world to a disturbing degree. If we turn these "refined methods and products" back to the "needs and problems out of which they arise and which they have to satisfy to be acknowledged," we find ourselves in a bit of a bind. Mazel argues that the world considered as constructed by discourse helps expose past wrongs and empowers oppressed peoples; while I wish to honor his concern, I am also convinced that considering the world solely a "socially constructed artifact" commits injustice to the physical world as a participant in human culture and also deserving of ethical consideration—consideration that will be essential to everyone's survival. In his essay "Environmentalism and Multiculturalism," Dan Flores concludes that "diversity in human culture just may be almost as important to adaptation and evolution on earth as we have long believed ecological diversity to be. In the best philosophical sense, choosing between humans and nature is a non sequitur" (36).

Criticisms that deny out of hand even the possibility of experiencing the physical world and that sustain the dualistic opposition of nature and culture will remain inadequate. Such a critical stance sets one beside experience, not along with it, and "forbids discourse the project of evoking the natural world through verbal surrogates and thereby attempting to bond the reader to the world as well as to discourse" (Buell, *Environmental Imagination* 102). Consequently, it forbids the attempt to bond the reader to experience, impeding criticism's opportunity to exist—as we all do, even literary critics—in that ordinary world where experience is often coarse, crude, and brutish, but also shot through with beauty and meaning. "Underneath the rhythm of every art and of every work of art, Dewey contends, "there lies, as a substratum in the depths of the subconscious, the basic pattern of the relations of the live creature to his environment" (*AE* 155).

In the following chapters I will examine how the ideas outlined in this introduction are played out in American ecological writing since 1911. The most prominent nodes I hope to bring to light are ones where the ideas of ecology and democracy grow closer together, are along with one another. Perhaps this is best illustrated by thinking that on the fringe of each of these ideas are barely conceived borders where an exchange between the two can happen—an ecotone. Again, an ecotone is the transitional zone between two or more ecosystems where evolutionary potential becomes most possible. It is a place of growth and contingency, a mediational space where change can happen. An ecotone exists between ecology and democracy. It can be a place of joy. Some of the nodes where the two ideas rub up against each other are the concept of interdependence; the notion of borders and barriers as permeable and transitional; and the need for public access to knowledge; further, acts of intelligence participate in nature and culture; experience is a cumulative process, with an emphasis on the input of everyday life; human culture is embedded in physical nature, and physical nature is embedded in human culture. There is an ongoing exchange of creative energy among these ideas that powers pragmatist ecology.

Dewey writes in *Art as Experience:*

> Since the ultimate cause of the union of form and matter in experience is the intimate relation of undergoing and doing in interaction of a live creature with the world of nature and man, the theories, which separate matter and form, have their ultimate source in neglect of this relation. Qualities are then treated as impressions made by things, and relations that supply meaning as either associations among impressions, or as something introduced by thought. There *are* enemies of the union of form and matter. But they proceed from our own limitations; they are not intrinsic. They spring from apathy, conceit, self-pity, tepidity, fear, convention, routine, from the factors that obstruct, deflect, and prevent vital interaction of the live creature with the environment in which he exists. (137–38, Dewey's emphasis)

We need both art and science for this vital interaction, to perceive the relations essential to meaningful experience. When Dewey excludes "impressions made by things," he means that not the things themselves but the relations among them are centrally important. Relations are not "associations"; association of species traits, for example, allows categorization but does not imply a mutual relationship. A relationship implies interdependence. And surely thought is activated in the creation of form, but thought does not instantiate relations. The relations are there in the world prior to our thinking about them, often just out of sight, and human beings and their arts and sciences become inextricable parts of these interrelationships. Experience arises from the perception of these relations. In *The Open Space of Democracy*, Terry Tempest Williams writes: "John Dewey in 1937 said, 'Unless democratic habits of thought and action are part of the fiber of a people, political democracy is insecure. It cannot stand in isolation. It must be buttressed by the presence of democratic methods in all social relationships'" (87).[13] Humans as a species cannot, any more than Dewey's individual, "stand and act alone," and through the perception of relations among different arts of knowing, and understanding the central importance of those relations to experience in a democratic community, we can create knowledge genuinely humane. We can live along with, not isolated from, one another and the physical world, and, following Dewey, our responsibility most emphatically entails understanding all problems, social and environmental, as remedial.

Hetch Hetchy Valley, 1911. Courtesy of the Library of Congress, LC-USZ62-135261.

1
An Arc of Discovery
John Muir's *My First Summer in the Sierra*

If the earth opens and swallows me up, this need not prove that my trust in it was misplaced. What better place for my trust could there be?
—Stanley Cavell

Those who like Muir, Emerson, and Thoreau, saw in nature a link with divinity, were not mistaken. All mythology and religion in one way or another acknowledges that primary relation, the very foundation of religious instinct. With continual alteration and settlement of the land, that connection, lacking an adequate cultural mediation, becomes more difficult to discover and maintain, but to the alert and sensitive spirit the secret of things remains intact.
—John Haines

After a long walk from Indiana to Cedar Key on the Gulf Coast of Florida, after a months-long recovery from malaria, and after a sea voyage to Cuba, then on to New York, then back aboard ship to Panama and San Francisco, John Muir wandered into the great Central Valley of California in 1868 and tended sheep for a time at Twenty Hill Hollow. He left Twenty Hill Hollow in the spring, and in the summer of 1869 the thirty-one-year-old Muir, a shepherd named Billy, a Chinese and a Native American helper, and Carlo the St. Bernard helped drive Patrick Delaney's flock of more than two thousand sheep to high pasture in the Sierra Nevada. This was Muir's first summer among the peaks of California, and more than forty years later, in 1911, *My First Summer in the Sierra* appeared, drawn from journals and notebooks and written when Muir was seventy-three. The journal-like entries are temporally structured by date and run from 3 June to 22 September 1869—the same year, incidentally, that Ernst Haeckel, a biologist at Friedrich Schiller University in Jena, Germany, coined the term *ökologie,* or ecology. The book is dedicated to the Sierra Club of California.[1]

My First Summer in the Sierra narrates the story of Muir's initial awakening to the power of the particular landscape that inspired the best of his writing and in which he spent some of the most meaningful moments of his life. William and Maymie Kimes insist that "this book, published near the apex of his career, reaps the competence of age while capturing the essence of youth, and becomes, we believe, his finest book" (85). Michael Cohen observes that the structure of *My First Summer* makes it possible that "The young narrator could travel his own innocent and sometimes heedless ways, while the older and wiser man constructed the bleak social context of the young man's travels. . . . The result is a carefully orchestrated structure which shows young Muir torn between his desire to wander free in the wilderness until he found a home and the necessities that required him to earn his bread and to encounter other men" (351). One of the central texts in the American nature writing tradition, *First Summer* has often been read as testimony to Muir's sudden conversion to nature seer during his first experience in the High Sierra—one could say the book has become eco-mythic.

However, the text raises some interesting questions. To complicate the above assessments, the original journals of the trip are lost, and *First Summer* seems to have been taken mostly from three extant handwritten notebooks dated 1887 that are either a revised, reworked, or totally rewritten version of the summer of 1869 (see Holmes 253–59).[2] In his biography of Muir, Steven Holmes has convincingly shown—by looking closely at the years leading up to Muir's arrival in California and underneath *First Summer* to the 1887 notebooks and to the journals surrounding that mythic year Muir first walked into the Sierra—that Muir's ecological vision of the world was developed over his lifetime, was not an epiphany. It was the same kind of process that Dewey describes in *Art as Experience*:

> The act of expression that constitutes a work of art is a construction in time, not an instantaneous emission. And this statement signifies a great deal more than that it takes time for the painter to transfer his imaginative conception to canvass and for the sculptor to complete his chipping of marble. It means that the expression of the self in and through a medium, constituting the work of art, is *itself* a prolonged interaction of something issuing from the self with objective conditions, a process in which both of them acquire a form and order they did not at first possess. (70–71, Dewey's emphasis)

The work of art as it becomes public is the culmination of a process—an "interaction." Interaction, of course, requires more than one participant, and Dewey designates that other participant as "objective conditions," which we can take to be both the cultural and physical environments. It is crucial to point out here that Dewey, with his choice of the term "objective," does not mean a static world absolutely independent of the self. The act of expression engages the world, contributes to ongoing experience. Referring to William James's notion of a "double-barreled word" (James, *Essays* 10), a word possessed of complex shades of meaning, Dewey insists that "'Experience' denotes the planted field, the sowed seeds, the reaped harvests, the changes of night and day, spring and autumn, wet and dry, heat and cold, that are observed, feared, longed for; it also denotes the one who plants and reaps, who works and rejoices, hopes, fears, plans, invokes magic or chemistry to aid him, who is downcast or triumphant. It is 'double-barrelled' in that it recognizes in its primary integrity no division between act and material, subject and object, but contains them both in an unanalyzed totality" (*EN* 18). The human and nonhuman, the cultural (magic, chemistry) and natural (changes of night and day), are interrelated; dualities dissolve, and Dewey emphasizes their dissolution by choosing examples from science, nature, and agriculture. Rejoicing in the harvest cannot happen without the change of seasons. Within experience, there is fluidity between subject and object. For Dewey, the split between subject and object, rooted in the distinction between human beings and the rest of nature, and insisted upon by Western culture since the time of the Greeks, is philosophical shorthand accepted as received truth that has remained unquestioned for far too long. Most of his work calls this duality into question. Both sides of the traditional split are part of the same process, existing in a complex web of relations. There is, of course, a sense of otherness. Dewey does not posit an identity between subject and object; however, he insists that in experience the self and the other influence one another to the extent that it makes little sense to think about them outside the context of experience. Through the work of art, which is itself a culmination of the artist's historical interaction with the physical and cultural worlds, change happens to both world and self. Art in this way engages ethics and politics. This is how Muir came to be changed by the Sierra, and it is safe to say that he hoped readers' perceptions of the physical world would come to be changed through experiencing his art. There are no epiphanies in this context.

About Muir's creative process, Holmes makes clear that we have no way of knowing for certain whether the 1887 notebooks are a faithful recording of the 1869 journals, though through comparison with the extant journals he makes a convincing argument that the notebooks had already been revised with an eye toward publication. This destabilizes the text to a large degree, and what we have is most likely a reliable record of facts, places, and landscapes, but with the older Muir's fully evolved ecological outlook attached to that experience (Holmes 259). We also need to keep in mind that at the time of its publication, Muir was at the height of his fight to preserve Hetch Hetchy Valley from being dammed, and it is not beyond question that the rhetoric of *First Summer* is intended to have a political effect on a public he wished to draw to his side of the fight. Daniel Philippon has investigated the discursive frames that environmental writers employ rhetorically to help create social change on behalf of the environment. For Muir, the overarching metaphor is nature as a park, and his strategy is to draw people into the Yosemite in order to convince them to love and defend it. Clearly, this is happening in *First Summer*. Muir uses the metaphor of the first visit along with that of a spiritual awakening to arouse political action.[3]

Hetch Hetchy Valley is a neighboring valley to the Yosemite, and, according to Muir, was nearly as marvelous. Hetch Hetchy was not included within Yosemite National Park when the latter was created in

1890, largely through the lobbying of Muir and Robert Underwood Johnson, although Hetch Hetchy was a designated wilderness preserve. As early as 1882 the valley had been eyed as a site for a possible dam to provide hydroelectric power and drinking water for San Francisco. San Francisco had once before applied for permission to turn the valley into a reservoir and was denied. However, in 1906 San Francisco suffered a great earthquake and fire, and public opinion on the matter shifted. The application was granted in 1908, and Johnson and Muir led a national campaign to save Hetch Hetchy Valley. The issues raised in that debate remain part of our current political climate, and as Roderick Nash writes in *Wilderness and the American Mind:* "for the first time in the American experience the competing claims of wilderness and civilization to a specific area received a thorough hearing before a national audience" (162). Muir lost his fight, and in 1913 President Woodrow Wilson signed Hetch Hetchy over to the city of San Francisco (it remains submerged to this day). Muir died a year later.[4] So *First Summer* is clearly more than a text about nature; through art, Muir engages democratic politics. Advocacy and art are linked in their urge to change the people's perception of the physical environment and their role in it.

We clearly have an experience that is mediated to a great extent, not only through the published text but also through journals, notebooks, and politics, all participating in both physical and cultural environments. Cohen thinks that at this late point in his life, Muir had resigned himself to the incompatibility of wilderness and civilization, but with *First Summer*'s focus on interrelationship and process, it seems to me that Muir may well have hoped that the textualization of his early experience would initiate political and ecological awareness of how the physical world and human culture function together. For twenty-first-century readers, Muir's perception of interrelatedness and its communication are certainly far more important than his paeans to wild nature (Cohen 358).

This is not to deny the power of that first summer, but Holmes rightly insists we acknowledge that Muir underwent a long development before he came to his most profound insights, and such acknowledgment recognizes Muir as an even more powerful cultural figure because beneath the myth of ecological prophet, useful as that myth may at times be, is the story of a real person attaining over time an extraordinary awareness of the world around him. "My sense is that Muir did not suddenly *find* a new home in Yosemite," writes Holmes, "but rather *made* one there over the course of years" (200–201, Holmes's emphasis). It is more difficult, surely, to learn from an epiphany than it is to learn from another's life experience; in other words, *First Summer* should be read not as a call to run off to the woods or mountains and expect to be reborn in a New Age way, but as a highly mediated text that contains Muir's attempt to represent a way of perceiving the world that had taken him a lifetime to accomplish. In other words, change happens in the course of experience. This is not to deny that, as Philippon comments, "Whether it happened over the course of a summer or over the course of several years, the spiritual change that took place in Muir remains significant" (128). Muir's way of perceiving the world led to a heightening of his ecological consciousness, which in turn contributed to his advocacy.

Holmes, of course, in order to get at the facts of Muir's biography and arrive at the historical Muir, was compelled to set aside much of Muir's rhapsodic bearing and ecological insight in *First Summer*. I suggest that *First Summer* in its finished form can provide a kind of guide as to how we might *make* a new home of the places in which we already *are* by replacing the categories Holmes brackets out in his attempt to re-create Muir's early life, and in so doing come to understand modes of perception discovered in Muir and the other writers in this study as possible strategies for ways we can live "along with" one another and the physical world. Ecological writing may help perform the cultural mediation Haines finds lacking in this chapter's second epigraph. All of this will take long effort, which is, of course, one of Holmes's most valuable insights.

Certainly, Muir worked within both a social discourse and a literary tradition, and in what follows I try to see his relation to cultural figures and conventions less as an anxiety of influence and more as an ecology of influence. Muir writes in his journal that "no amount of word making will ever make a single soul to *know* these mountains" (Wolfe, *Journals* 95), and if he believed his own words, he went on to make a lot of them in spite of himself. Words that encourage human beings to perceive the natural world as an active participant in texts and in experience could have a profound effect on the structures within which we currently live. Special attention to the interconnections among living and nonliving things can recuperate a sense of responsibility to the world and its inhabitants. I will pay heed to Muir's habit of close attention to natural detail and to his representation of physical processes in his attempt to make others *know* the Sierra. But the vitality of Muir's work is not contained by the mountain range. In his work, Muir represents the entire human and nonhuman world as part of a community of beings lodged in a physical world just outside, or on the fringe, of human focus and expression. It is often at these fringes or edges—these ecotones—that the perception of relations occurs. Also, Muir often hovers between Romantic and Victorian ways of perceiving the natural world, but like his predecessor Thoreau, Muir ultimately takes up a counterposition that insists upon an ecological understanding of experience in which the world and human beings participate. In other words, Muir helps move his readers into a more modern understanding of the world as process. Because of his status as a transitional figure beginning in the nineteenth century and ending on the brink of modernism, and because of his opening nationwide public discussion and advocacy of the environment, Muir provides a baseline for the rest of the study.

As a seminal presence in the formation of the preservation movement in the United States, and as a founder of the Sierra Club, Muir remains a powerful voice in the environmental movement. This is partly because he was in a position to dissent significantly from the dominant conservation ethic of his time and ours, voiced most influentially by Gifford Pinchot as "the development and use of the earth and all its resources for the enduring good of men" (qtd. in Worster 266).[5] Pinchot and Muir institute the split between conservationists and preservationists that still exists in the United States. Inverting the sense of Pinchot's comment in the text of *First Summer*, Muir writes: "The basin of this famous Yosemite stream is extremely rocky,—seems fairly to be paved with domes like a street with big cobblestones. I wonder if I shall ever be able to explore it. It draws me so strongly, I would make any sacrifice to read its lessons. I thank God for this glimpse of it. The charms of these mountains are beyond all common reason, unexplainable and mysterious as life itself" (258–59). Here, metaphor seems to displace the real nature of Yosemite, figuring it as a street, a man-made object. However, a street is also a thing along which human beings move toward a destination, and because Yosemite Valley is clearly the path along which Muir found his life's calling, the metaphor seems on one level apt. The natural world does not mirror Muir's mind back to itself in a high Romantic way, and Muir shifts his focus toward the meaning content of the landscape itself.[6] Contra Pinchot, Muir sees meaning lodged in the landscape, not only in its relation to his own discourse, although that relation is also centrally important. The art of knowing depends upon the emergence of meaning from the interrelationship of environment and creative intelligence.

At the same time Muir thanks God for a glimpse of meaning, he levels the presence of the interpreting self by asserting its inability to read. The figure of nature as a book also marks Muir's indebtedness to Louis Agassiz, whose *Études sur les glaciers* greatly influenced Muir, and Muir's inability to read the text also indicates a shift, this time away from Agassiz's pre-Darwinian view of nature. By the time Muir wrote *First Summer* Darwin's influence was ubiquitous, and like his movement along a spectrum away from Romanticism, Muir's scientific views begin to wander from the idea of nature as a decipherable text to nature as a flowing process, although, of course, the residual effects of Darwin's early influences never vanish.[7] Darwin's influence was also instrumental in Dewey's understanding of experience as a process: "Old questions are solved by disappearing, evaporating, while new questions corresponding to the changed attitude of endeavor and preference take their place. Doubtless the greatest dissolvent in contemporary thought of old questions, the greatest precipitant of new methods, new intentions, new problems, is the one effected by the scientific revolution that found its climax in the *Origin of Species*" ("Darwinism" 14). Attitudes of endeavor were indeed changing, and St. Armand comments that the metaphor of nature as a book was moribund by the time Muir had access to it. He notices a drift in Muir's metaphors from nature as book to nature as palimpsest, and this shift in metaphor parallels the erasure and rewriting of old thought by contemporary thought that marks Dewey's debt to Darwin (St. Armand 38). In one sense Darwin is to Muir and Dewey as Virgil is to Dante: a guide through a volatile, unstable world. The tension generated by this instability is evident in Muir's willingness in this passage to "make any sacrifice," to negate the self, in order to interpret the book of nature, "a grand page of mountain manuscript that I would gladly give my life to read" (*First Summer* 102), but a page just out of his reach.

Muir seems caught in a paradoxical situation when faced with the desire to both read the landscape and efface the interpreting self in response to that same landscape. This situation points to the continuing problem of the self in ecological writing. The form itself seems to demand an often microscopic attention to natural detail, which seems to insist on extraordinary powers of perception on the part of the narrating self, yet that self must fall away in order to grant participatory status to the natural phenomena it describes so strenuously if it wants to engage an ecological relationship. This "aesthetics of relinquishment," as Buell calls it, demands not the "eradication" of self but rather the "suspension of ego to the point of feeling the environment to be at least as worthy of attention as oneself and of experiencing oneself as situated among many interacting presences" (*Environmental Imagination* 178). Buell's argument echoes one Dewey made sixty years earlier in *Art as Experience:* "Where egoism is not made the measure of reality and value, we are citizens of this vast world beyond ourselves, and any intense realization of its presence with and in us brings a peculiarly satisfying sense of unity in itself and with ourselves" (199). The sense of unity—often temporary—arises from a realization that the self participates in the world, marked by Dewey's repetition of the terms "with" and "in" to delineate the relations between the human and the world beyond us. Dewey intuits a sense of interrelationship between us and the world, and the reproduction of details heightens the sense of the landscape's presence at the same time that the author levels the presence of the self in the text; again, this seems to me to be a profoundly ecological idea.[8] Egotism, or even anthropocentrism, is no longer the whole measure of reality. The point is, finally, less to evoke the physical world at the expense of the human than to forefront the idea that both participate in experience. To diminish either, then, is to diminish potential experience. The strategy displaces the human self from a position of omniscience and posits it as a participant in experience.

In a famous section of his first book, *The Mountains of California* (1894), Muir discloses just how destabilizing a mountain—a particular landscape—can be. In chapter 4, "A Near View of the High Sierra," Muir leads a group of artists high into the mountains in search of a paintable vista, protesting that the Sierra peaks cannot be contained by powerful aesthetic categories such as the picturesque, as "few portions of the High Sierra are, strictly speaking, picturesque. The whole massive uplift of the range is one great picture, not clearly divisible into smaller ones"; the range cannot be separated into "artistic bits capable of being made into warm, sympathetic, lovable pictures with appreciable humanity in them" (*Mountains* 38–39). The effort to temper the human presence in the landscape is evident at this point. Rebecca Solnit traces this effect of the landscape painting tradition on the formation of the national parks generally and on Muir in particular. Rather than seeing the minimizing of human presence in the landscape as an ecological turn, she sees it as an erasure and obliteration of the real lives led in particular landscapes and ecologies.[9] She writes that "perhaps our presumed alienation from nature cannot be alleviated by scenery, perhaps it requires a more profound engagement with the natural world as a system in which we are enmeshed, which feeds us and takes our wastes" (266). Although Muir at times derides the human presence in the landscape, I am not willing to dismiss the possibility that his is also a strategy for including or inserting the human being more fully into ecological systems, just as Solnit rightly insists we do. Philippon is insightful here too: "For Muir, of course, such a seeming 'reduction' in human status was nothing of the sort; if all creation was sacred, then humankind shared in that sacred status, no worse off than any other plant or animal, though no more privileged either" (145). I am also drawn back to Thoreau's sandbank passage, where the narrating self is submerged yet remains very much enmeshed in a landscape that does figuratively take animal waste. In Muir, the living human body at times does indeed "blend into the rest of nature, blind to the boundaries of individuals" (Wolfe, *Journals* 78), but the remains of the dead human body are likewise recycled into the earth, turned from waste into life: "Instead of the sympathy, the friendly union, of life and death so apparent in Nature, we are taught that death is an accident, a deplorable punishment for the oldest sin, the arch-enemy of life, etc." (Muir, *Thousand-Mile Walk* 70). In both Thoreau's and Muir's terms, the landscape is far more than scenery: it is an active agent in biological processes.

Additionally, Muir's debt to landscape painting contains ecological possibility. He continues: "Climbing higher, higher, new beauty came streaming on the sight: painted meadows, late-blooming gardens, peaks of rare architecture, lakes here and there, shining like silver, and glimpses of the forested middle region and the yellow lowlands far in the west" (*Mountains* 43). This view is nothing so much as a conventional picture, except for Muir's interesting use of the word "streaming." Beauty here is not a static entity, but flowing. It moves. Then, from the top of the divide, the scene becomes less conventional: "Beyond the range I saw the so-called Mono Desert, lying dreamily silent in thick purple light—a desert of heavy sun-glare beheld from a desert of ice-burnished granite" (43). From this vantage point Muir shows the reader an extraordinary, gleaming world of heat, ice, and rock, and Muir's near sacral use of light reminds one of American luminist paintings. His balancing act between deserts suggests both the "equipoise" of luminist paintings and the balancing of the artist's ego with the landscape. "Such paintings, in eliminating any reminders of the artist's intermediary presence," writes Barbara Novak, "remove him even from his role of interpreter" (44) in a painterly version of the writer's respectful engagement with the landscape.

His role as interpreter diminishes as Muir enters an unfamiliar landscape on his approach to Mt. Ritter, and the short prose exposition that follows is on one level simply a terrific climbing narrative. But it reaches far deeper than that, and, like Dewey's description of experience, it "thus reaches down into nature; it has depth (*EN* 13). Cohen cogently argues that "An unknown region in his [Muir's] consciousness had been awakened in this unknown region on Mt. Ritter" (69). Muir recalls that at around 12,800 feet above sea level, about halfway up the cliff face, "I was suddenly brought to a dead stop, with arms outspread, clinging close to the face of the rock, unable to move hand or foot either up or down. My doom appeared fixed. I *must* fall. There would be a moment of bewilderment, and then a lifeless rumble down the one general precipice to the glacier below" (*Mountains* 51, Muir's emphasis). Face to face, Muir is plastered to rock, and the destruction of the self seems certain.

His paralysis temporary, change of self occurs in the face of the seeming necessity marked by Muir's italicized *"must."* Life asserts itself in the face of fear, but in this experience its assertion seems beyond self-will:

> When this final danger flashed upon me, I became nerve-shaken for the first time since setting foot on the mountains, and my mind seemed to fill with a stifling smoke. But this terrible eclipse lasted only a moment, when life blazed forth again with preternatural clearness. I seemed suddenly to become possessed of a new sense. The other self, bygone experiences, Instinct, or Guardian Angel,—call it what you will,—came forward and assumed control. Then my trembling muscles became firm again, every rift and flaw in the rock was seen as through a microscope, and my limbs moved with a positiveness and precision with which I seemed to have nothing at all to do. Had I been borne aloft upon wings, my deliverance could not have been more complete. (51–52)

Cohen in his reading of this passage argues that Muir decided on casting this narrative in the passive voice because "he wished to suggest a kind of dual consciousness. The climb happened to him. He could only observe the phenomenon" (68). But there is a sense here too that the self has relinquished control to a larger degree than Cohen suggests: Muir's mind shifts from the readily accessible cultural figure of "Guardian Angel" to geological fact, to "every rift and flaw in the rock," and then "I seemed to have nothing at all to do." On Mt. Ritter he saw into the rock and descried the texture of the crystals themselves. It is as if he looks at the world through a microscope, and through his focus on the rock, the self recedes and something else surfaces, crystallizes, guiding Muir upward toward safety. Only after he loses himself in the structure of the rock does he find the wherewithal to save himself, yet the human self remains displaced from a position of control. As the duality between self and world comes under erasure, the self evolves into a participant in a transformational experience that includes the rock itself as a primary agent.

Thus the entire experience on Mt. Ritter might also be seen as a figure for Muir's method of writing. In *First Summer,* Muir writes: "How deeply with beauty is beauty overlaid! The ground covered with crystals, the crystals with mosses and lichens and low-spreading grasses and flowers, these with larger plants leaf over leaf with ever-changing color and form, the broad palms of the firs outspread over these, the azure dome over all like a bell flower, and star above star" (128). His prose reveals a Victorian preference for cramming as much as possible into the sentence, and here it moves in an interesting way from the smallest particles, rock crystals, through an interwoven world of stunning beauty. From the soil to the solar system, things overlap, and these objects remain separate, yet are all involved in one enormous, fluid system. Muir saw this world-system as an aesthetic process, whereas we may be more likely to call the process ecology. I suggest it is both at the same time.

Toward the end of *First Summer,* Muir's vision encompasses the interrelationships among things:

> Contemplating the lace-like fabric of streams outspread over the mountains, we are reminded that everything is flowing—going somewhere, animals and so-called lifeless rocks as well as water. Thus the snow flows fast or slow in grand beauty-making glaciers and avalanches; the air in majestic floods carrying minerals, plant leaves, seeds, spores, with streams of music and fragrance; water streams carrying rocks both in solution and in the form of mud particles, sand, pebbles, and boulders. Rocks flow from volcanoes like water from springs, and animals flock together and flow in currents modified by stepping, leaping, gliding, flying, swimming, etc. While the stars go streaming through space pulsed on and on forever like blood globules in Nature's warm heart. (236)

Obviously this is an ecology in motion, and the fluid nature of Muir's prose, emphasized by the abundant use of present participles and the extended metaphor of stream flow, makes no distinction as to whether the flowing substance is water, ice, rock, mammal, bird, or fish, and it recalls Thoreau's sandbank passage as the metaphor shifts to the circulatory system of nature. The other metaphor for the universe here is the "lace-like fabric" of these braided streams, and universal processes are all interrelated like the intricate needlework of a piece of Holland lace—of a beautiful thing. Glaciers, avalanches, and, by extension, the rest of the processes named here are "beauty-making," and that beauty is clearly dependent on the interrelationship of the processes named here. The passage is finally subsumed in "Nature's warm heart." It is incorporated into the whole, which Muir names "Nature."

In order to help see how this interrelationship is consummated as aesthetic experience, or how ecology might be analogous to that experience, I want to move briefly to Dewey and then to Neil Evernden. For Dewey, one of the ramifications of aesthetic experience is cooperation or participatory relations between the human and physical world: "For the uniquely distinguishing feature of aesthetic experience is exactly the fact that no such distinction of self and object exists in it, since it is esthetic in the degree in which organism and environment cooperate to institute an experience in which the two are so fully integrated that each disappears" (*AE* 254). Although this full integration happens only temporarily—it is fleeting like perchings and flights—this consummatory moment seems analogous to what Muir understands as "Nature's warm heart." Although the narrating self is submerged in the process of the flow of nature, it also participates as the creative intelligence behind the work of art itself. The sense that human beings are participants in a larger world around us seems always on the fringes of our awareness, and it is only by bringing intelligence to bear on the relations we just barely perceive that aesthetic experience is possible.

Evernden complicates this notion for us. He points out that chloroplasts in plants and organelles such as mitochondria in humans, which were once thought to be parts of cells, are quite distinct from the cells in which they are found. Mitochondria move from cell to cell and replicate independently of a cell. There are extra-chromosomal structures in mammalian bodies that can move from cell to cell. The implication, as Evernden says, is that "it is conceivable that groups of species, perhaps whole communities of organisms, could, in a sense, co-evolve. They are quite literally interrelated" (40). He then asks a series of questions: "Where do we draw the line between one creature and another? Where does one stop and the other begin? Is there even a boundary between you and the non-living world, or will the atoms of this page be part of *you* tomorrow? In short, how can you make any sense out of man as a discrete entity?" (40).[10] Dewey in some rather garbled prose anticipates Evernden when he writes about intradermal transactions. The organism is not necessarily bounded by its dermal layer, and "[t]he anticipated future development of transdermally transactional treatment has, of course, been forecast by the descriptive spade-work of the ecologies, which have already gone far enough to speak freely of the evolution of the habitat of an organism as well as the evolution of the organism itself" (*KK* 117). This is where the science of ecology becomes subversive: What does it mean for human beings to fully conceive of themselves as integral parts of the world, involved and evolving not as a self removed from the natural world but as contributing parts of a larger whole—of an earth-wide lacework or patchwork of ecosystems, as potentially beauty overlaid upon beauty? This is, of course, a difficult question, but it can afford us the imaginative space to bring our intelligence to bear on relations, to activate the art of knowing to create mutually beneficial relationships with the rest of creation. What cellular biology now tells us, Muir seemed to have intuited in the passage above which figures the world as one large, beautiful life process. In his aesthetics, Dewey, too, shows an uncanny awareness of our implication in the world: "The epidermis is only in the most superficial way an indication of where an organism ends and its environment begins" (*AE* 64). The creature needs this interconnection to its environment in order to create art, to experience in an aesthetic consummatory way, a way I would like to think of as ecological.

In a pragmatist ecology, the world and the observer both participate in experience through intense concentration and patterns of observation. By focusing so intently, there is a push and shove along the ecotonal boundary of self and world until the two seem to function together, heightening experiential possibility. The way Muir functions back there on the face of Mt. Ritter and the way he reconstructs the experience in prose are, again, remarkably close to Dewey's exposition of having an experience. In chapter 3 of *Art as Experience,* "Having an Experience," Dewey writes: "Because of continuous merging, there are no holes, mechanical junctions, and dead centers when we have *an* experience. There are pauses, places of rest, but they punctuate and define the quality of movement. They sum up what has been undergone and prevent its dissipation and idle evaporation. Continued acceleration is breathless and prevents parts from gaining distinction. In a work of art, different acts, episodes, occurrences melt and fuse into unity, and yet do not disappear and lose their own character as they do so" (43, Dewey's emphasis). Muir's position stuck on Mt. Ritter's rock face is a pause, a punctuation mark, helping to define the quality of his experience. From the moment "life blazed forth" there is breathless acceleration up the face of the mountain, and the reconstruction of the experience in prose implies a unity in which each part—human and mineral, muscle and rock—maintains its integrity in motion.

As Dewey suggests, a work of art emerges from ongoing experience, and integral to that experience is the "interaction of organism with its environment, an environment that is human as well as physical" (*AE* 251). In an ecological sense, this experience can be seen as ecosystemic—it is dependent upon doing, undergoing, interaction, reaction, and response in a kind of feedback loop. It is a relationship through the porous boundary of self and world. Millions of these relationships take place at once, inhabiting a larger whole of various dimensions, from a mountain range to a tide pool to a film of water around a grain of sand. The whole remains viable due to the unique contributions of integral members, or, as Muir more simply said, "When we try to pick out anything by itself, we find it hitched to everything else in the universe" (*First Summer* 157).

This conflates part and pattern, and distinctions blur. We are never only a part or only a pattern, but severally patterns and parts, just like light is never only a wave or a particle. Muir functions in patterns of nearness (on Mt. Ritter) while concurrently hitched to the vastness of the universe. He is capable of the near and the far view. In his entry for 26 August, Muir writes, "Frost this morning; all the meadow grass and some of the pine needles sparkling with irised crystals,—flowers of light" (*First Summer* 234). He looks closely enough to see that some, not all, of the pine needles are sheathed in ice. The crystals are "irised," so they in a sense can return Muir's glance. His prose expands from the grass and pine needles outward into the realm of light. This light comes into presence through the crystals of frost on grass and needles; his prose moves "through the parts to enter into the whole which becomes present within the parts" (Bortoft 12). For an ecological aesthetics generally, or for a participatory democracy, there is no permanent separation of parts from the whole, and the predominant quality of a pragmatist ecology is the introduction of a participating world into the art of human beings. Pragmatist ecology advocates an aesthetics of interrelationship, and the perception of beauty as both natural connection and culturally constructed category can permeate human experience, thus changing how we understand our role in our communities and environments. Thought becomes active, and as Richard Shusterman argues for the intervention of pragmatist aesthetics into contemporary art, "Since art is a crucial instance and cherished resource of human flourishing, philosophy betrays its mission if it merely looks on with abandoning neutrality at art's evolving history without joining the struggle to improve its future" (45).

Our future may well depend upon our looking closely enough, being attentive enough, so that it becomes difficult to miss our connections to the natural world and becomes possible to re-see connections in our art that have been overlooked—for instance, the ecological possibility mentioned in reference to landscape painting. Ecological writing seeks to exercise this aesthetics. It is safe to assume that different people will perceive these interrelationships more effectively in different art forms and landforms. Different people and art flourish in different places. The alpine environment, from its vast panoramas to its ice-encased pine needles to the kitchen middens of Douglas squirrels, was the ideal place for Muir, and certainly his childhood experiences of and resistance to enclosure contributed to his sense of aesthetic expansion in the mountains.[11] But most important for my thinking is the democratic notion that one can perceive the aesthetics of interrelationship anywhere—it is a too-often-overlooked condition of our everyday world and our everyday experience.

Both Emerson and Thoreau advocated for the importance of the local for philosophical endeavor, and Muir's deep indebtedness to Thoreau is best documented by Richard F. Fleck.[12] For example, Muir's essay "Twenty Hill Hollow" appeared in the July 1872 issue of *Overland Monthly*, and William Frederic Badè appended an edited version of the essay to *A Thousand-Mile Walk to the Gulf* to give continuity to the published material documenting Muir's life. "Twenty Hill Hollow" bridges a gap between the end of *A Thousand-Mile Walk to the Gulf* and the beginning of *First Summer*, and one of its most interesting aspects is that it allows us to track Muir in two passages as he works through the influence of Thoreau and Emerson. One of the passages Badè expunged is an almost-verbatim echo of the opening to Thoreau's famous essay "Walking." Muir writes, "I wish to say a word for the great central plain of California," whereas Thoreau in "Walking" wished "to speak a word for Nature."[13] Muir, following more closely the Thoreau of *Walden,* wished to say a word for a particular landscape, this time the landscape of California. The second passage, near the end of "Twenty Hill Hollow," is an inversion of Emerson, who wrote in *Nature:* "The currents of the Universal Being circulate through me; I am part or parcel of God" (10). At the end of his essay, Muir writes, "Presently you lose consciousness of your own separate existence: you blend with the landscape, and become part and parcel of nature" (86). Notice that in Emerson's version, the Universal Being is focused in the individual human being. Not so in Muir; he has reversed the relationship—the individual assimilates into the landscape—and thus the emphasis shifts from an egocentric understanding of the universal to an ecocentric understanding of a particular landscape. In Muir's essay, God becomes nature, perceivable in the California landscape.

In 1871, when Emerson, by then an old man, visited Yosemite, Muir was initially awed by this most influential American cultural figure. But Muir was also disillusioned. He was irritated and disappointed that Emerson would not camp out with him and that Emerson's entourage kept him from Muir's own temples of learning, the rocky heights of the Sierra, which to Muir dimmed the brick temples of New England.[14] Although one can easily understand why the aged Emerson would decline a camping trip with the effervescent young Muir, still, for Muir "both Emerson and Thoreau seemed insufficiently wild" (Fox 83). Although Muir thought Emerson "the most serene, majestic, sequoia-like soul I ever met" (Wolfe, *Journals* 436), and Emerson thought Muir "the right man in the right place" (Badè 259), Muir still felt that Emerson had not spent enough time out-of-doors.[15] It seems Muir wanted Emerson both "wild" and "serene." On his return to Concord, Emerson for his part sent copies of his essays, and Muir was, of course, honored to receive them; clearly, Emerson had a lasting influence on him. When Emerson left Yosemite to return to the East, Muir wrote: "though lonesome for the first time in these forests, I quickly took heart again—the trees had not gone to Boston, nor the birds; and as I sat by the fire, Emerson was still with me in spirit, though I never again saw him in the flesh" (*National Parks* 136). After visiting Emerson's and Thoreau's gravesites years later, Muir wrote to his wife, "I did not imagine I would be so moved at sight of the resting places of these grand men as I found I was, and I could not help thinking how glad I would be to feel sure that I would also rest here" (qtd. in Gifford, *Muir* 311).

However, as Stephen Fox emphatically points out, Muir did not read Emerson with an adoring eye; quite the contrary, he interrogates and talks back to Emerson in the margins of the same books Emerson sent him:

> Emerson: "The squirrel hoards nuts, and the bee gathers honey, without knowing what they do." Muir: *How do we know this.*
> Nature "takes no thought for the morrow." *Are not buds and seeds thought for the morrow.*
> "It never troubles the sun that some of his rays fall wide and vain into ungrateful space, and only a small part on the reflecting planet." *How do we know that space is ungrateful.*
> "The soul that ascends to the great God is plain and true; has no rose-color." *Why not? God's sky has rose color and so has his flower. . . .*
> "The trees are imperfect men, and seem to bemoan their imprisonment, rooted in the ground." *No.* (qtd. in Fox 6)

Muir enters into a dialogic relationship with Emerson and blatantly refutes much of Emerson's transcendentalism. He says simply "No" to the ascendancy of human thought over the natural world. Of course, an Emersonian valorization of the human imagination had become increasingly difficult to sustain in the face of scientific developments inaugurated by scientists such as Lyell and Darwin, whose *Principles of Geology* (1830–33) and *On the Origin of Species by Means of Natural Selection* (1859) set human beings in an ongoing process—in the context of a vast depth of geologic time and in the context of evolution.[16] Darwin's influence on philosophy was also palpable. The art of knowing is also a process. In 1910, just one year prior to the publication of *My First Summer in the Sierra*, Dewey writes: "The influence of Darwin upon philosophy resides in his having conquered the phenomena of life for the principles of transition, and thereby freed the new logic for application to mind and morals and life. When he said of species what Galileo had said of the earth, *e pur si muove*, he emancipated, once for all, genetic and experimental ideas as an organon of asking questions and looking for explanations" ("Darwinism" 7–8). Of course, all things do move, and both Dewey's and Muir's sense of the world is a world in motion, unity always in transition. The environing world is never static, and Darwin's perception of this flux has "emancipated" thought. Dewey's use of the term "organon" also moves the body and the mind together. According to the *OED*, an organon is "a bodily organ, especially as an instrument of the soul or mind," and "an instrument of thought or knowledge: a means by which some process of reasoning, discovery, etc. is carried on." Body and mind are a continuum, and both body and mind together put the art of knowing into motion. Neither, according to Dewey, is the servant of the other. The two cannot be split apart for any reason, and "This no more occurs for the sake of the end than a mountain exists for the sake of the peak which is its end" (*EN* 84). In the moving, living world, natural and cultural, biological and moral relations exist in perpetual transition, and this transition, this contingency, is the prerequisite for questioning and explaining, for experience itself. It thus became increasingly difficult to subscribe to ideas that situate the human imagination, or soul, in an imperialistic posture which voices the imperative that "Within man is the soul of the whole" (Emerson, "The Over-Soul" 131). Although, sadly, this posture has never disappeared, both Muir and Dewey move toward a modern ecological view of the creative imagination as participating in, not lording over, the physical environment.

Like the journals of Thoreau before him, which "saved him from the excesses of the imperial imagination," Muir's journals-turned-book in *First Summer* provide "a context in which fact and idea could order themselves laterally, in a relation of relative equality, rather than hierarchically" (Peck 87). Muir looks closely at a pine cone, "cylindrical, slightly tapered at the end and rounded at the base. Found one to-day nearly twenty-four inches long and six in diameter, the scales being open" (*First Summer* 50), and then widens his perceptual focus to transform the cone-bearing tree into something more: "I never weary of gazing at its grand tassel cones, its perfectly round bole one hundred feet or more without a limb, the fine purplish color of its bark, and its magnificent outsweeping, down-curving feathery arms forming a crown always bold and striking and exhilarating" (51). Intense concentration on the facts of the cone and the tree develops through Muir's prose into the general qualities Muir finds so compelling in the forest of giant trees, and he moves back from the macro to the micro view, regretting that he "cannot draw every needle" (51). Muir's book constellates facts and ideas ecologically. Muir rejects Emerson's subordination of the natural world to human structures, insisting that "Nature is not a mirror for the moods of the mind" (Wolfe, *Journals* 190). As in Thoreau, there is instead a push and shove between the human and the nonhuman world.

Both Thoreau and Muir existed in a sometimes uneasy relationship to Emerson, and both seemed to exist likewise in a rather nettlesome way with John Ruskin, the most influential aesthetician of the late nineteenth century. About *Modern Painters,* Thoreau writes: "I am disappointed in not finding it a more out-of-door book, for I have heard that such was its character, but its title might have warned me. He does not describe Nature as nature, but as Turner painted her, and though the work betrays that he has given a close attention to Nature, it appears to have been with an artist's and critic's design. How much is written about Nature as she is, and chiefly concerns us, *i.e.* how much prose, how little poetry!" (*Journal* 1199). Thoreau should have known—it is about painters, after all—but his mistake is a reflection of just how strong Ruskin's impact was in the general culture even before his book was read. But also in this passage is Thoreau's desire to cut through cultural constructions of the natural world and see it as it pertains to everyday lived experience, and its pertinence does not exclude its written mediation; in fact, nature as it participates in experience is poetry, is art ecologically conceived. Although it would be difficult to overestimate Ruskin's influence on Muir, Muir nonetheless took exception to Ruskin's moralistic, Manichaean obsessions as he saw them in chapters 19 and 20, "The Mountain Gloom" and "The Mountain Glory," of volume 4 of *Modern Painters*.[17] Echoing his complaint about Emerson, Muir insists that if Ruskin were "to dwell awhile among the powers of these mountains, he would forget all dictionary differences between the clean and the unclean and he would lose all memory and meaning of the diabolical, sin-begotten word *foulness*" (qtd. in Cohen 39–40). Although Muir resisted the dualism he perceived in Ruskin, Ruskin's insistence on a detailed understanding and faithful observation of the natural world as one of the highest achievements to which an artist can aspire is a demand Muir strove to meet for the rest of his writing life.

Roger B. Stein argues that in the nineteenth-century United States, "The Ruskinizing tendency seems to have been endemic to the Western explorers when they sat down to explain what they had seen." It led them "into a lyricism where the perceiving consciousness of the observer played as large a part" as the physical facts the writer sought to describe. However, Stein also points out that "by the 1880's his [Ruskin's] position as the expositor of nature had been almost completely undermined in the face of an increasing professionalization of the sciences" (166–67). Muir was one of the "latter-day transcendental naturalists" who participated in this critique of Ruskin, but Stein also claims that "Muir's Emersonian optimism was a late flowering of a spirit that was dying out of the American attitude" (166). We have already seen that Muir's was no unqualified acceptance of Emersonian optimism, although Muir struggled his entire life to hold onto his own, if not optimism, then enthusiasm. For Muir, this enthusiasm is not a gush of feeling but a summoning of energy pitched "at a responsive key in order to *take* in" (*AE* 60, Dewey's emphasis) that enables him to move beyond Ruskin's polarities and enter into a more modern experiential exchange with the environment that results in Muir's characteristic process-oriented environmental aesthetic.

Donald Wesling, like Frank Stewart and other commentators, sees a direct relationship between Muir and Ruskin: "Muir took his premise and method from Ruskin, sharing with the English writer a hope that, even in a technological culture, an implicative description might relate our sense of fact to our sense of value" (42). By "implicative description" Wesling means to closely read the details of the earth while displacing the self from the center of discourse in order to see "the earth as in itself the ultimate satire upon technology and greed" (40). At the end of *First Summer*, Muir writes that "The most telling thing learned in these mountain excursions is the influence of cleavage joints on the features sculptured from the general mass of the range," and the wild landscapes "look human and radiate spiritual beauty, divine thought, however covered and concealed by rock and snow" (254). A close observation of the geology of the terrain yields insight into spirit and divinity, beauty and thought, and the experience of Muir's initial summer cast into "implicative" prose shows the literary, environmental imagination in the process of creating a work of art in which author and landscape interrelate. The linkage here of geology, "cleavage joints," and the human and the spiritual is not an elision of the self. Muir tries to highlight the rock and snow as participants in an aesthetic process. In spite of this, like Stein in 1967, Wesling in 1977 seems to see Muir as a bit moribund: Muir was "a writer of the now defunct genre of descriptive prose" (43). I do not see Muir existing in as easy a relationship with Ruskin as Wesling's article suggests, and I have only one other issue to take with Wesling's essay: I do not accept his premise that descriptive prose is "defunct." The fact that a critic such as Dana Phillips was still, in 2003, trying to declare it defunct testifies to its resilience.

On the contrary, it is quite alive, and attracting more and more attention, as this project modestly proposes. Escalating environmental destruction on a global scale has played a role in sparking an environmentally conscious readership that desires close description of the natural world and clearly understands this writing and the experience it hopes to initiate as an intervention into the dominant discourse, which has allowed some of us to prosper at the cost of many millions of human others and at the cost of the physical world's continued integrity. Perhaps those who need such an intervention most are the students who will have to live and work in a world my generation has continued to degrade beyond reason. Certainly scholarship and pedagogy have a responsibility to those students, and we need to rediscover the role of the public intellectual epitomized by Dewey. Muir too lived a public life that traced an arc of discovery, and *First Summer* traces an arc of time. He retells his initial experience of discovering the Yosemite Valley and the mountains he called the "Range of Light" (236), and counter to a progress-of-civilization ideology, Muir revisits his capacity for wildness that he imagined could provide both a benefit for society at large and an antidote to the rapacious abuse of wild areas sanctified by the dominant ethics of his, and our, culture. At this late date it is more profitable to try to understand the possible cultural consequences his ecological perception of the natural world could have for us than to rely on the idea of the wild as sanctuary from culture.

As a public figure, Muir situated himself and functioned within a physical and cultural ecotone. He retells his experience in the Sierra, an alpine ecology rich in transition zones. He looks at his experience from the perspective of age, thereby working in an ecotone that blends past and present. Include us as readers, and we add the future dimension. And finally, Muir and his writing exist in a cultural ecotone undergoing dramatic change. He is influenced by changing theories and practices of science, industry, religion, and aesthetics; he embodies in himself and in his writing an ecotonal relationship between Romantic and Victorian ways of seeing the world, even as he moves beyond these categories toward a modern biocentric worldview. He is somewhere along a cultural arc that leads right to us in both our turn-of-the-millennium malaise and in our search for a viable relationship with the nonhuman world as we stumble into the twenty-first century.

What Muir found in and around Yosemite Valley was not a foundation but rather a shifting process enacted between a human being and the natural world. Toward the beginning of *First Summer,* he writes: "We are now in the mountains and they are in us, kindling enthusiasm, making every nerve quiver, filling every pore and cell of us. Our flesh-and-bone tabernacle seems transparent as glass to the beauty about us, as if truly an inseparable part of it, thrilling with the air and trees, streams and rocks, in the waves of the sun,—a part of all nature, neither old nor young, sick nor well, but immortal" (15–16). Muir imagines the human creature as inseparable from the beauty that surrounds it, but he is careful also to mark that humans are other from the nonhuman entities surrounding them through his use of the words "seems" and "as if." His prose functions to convince the reader of the possibility of capturing this elevated state of being, but it also resists the threat of complete dissolution of the self into the world. The human being and the physical world remain involved in the doing and undergoing of experience. Here is, of course, the wilderness prophet with an echo of Emerson's transparent eyeball, and with a somewhat inflated claim to immortality in which we can sense lingering traces of transcendentalist thought and, indeed, of the whole Romantic and Victorian tradition. But on the other hand, if we accept that people die and decay and feed microorganisms which in turn help build soil that nourishes plant life which feeds other human beings, then there is more than a little truth in what Muir says. He writes: "But let a child walk with Nature, let him behold the beautiful blendings and communions of death and life, their joyous inseparable unity as taught in woods and meadows, plains and mountains and streams of our blessed star" (*Thousand-Mile Walk* 70–71). We are all part of a cyclical process.

As are the Sierra peaks Muir returns to a few sentences later: "How near they seem and how clear their outlines on the blue air, or rather *in* the blue air" (*First Summer* 16, Muir's emphasis). Correcting himself midstatement, Muir revises his observation that the mountains are *on* the air, like a picture on a blue canvas. They are *in* the air, exist *in* a world process, not *on* a static background. Dewey seems to concur, perhaps to move one step further: "They *are* the earth in one of its manifest operations," part of a process (*AE* 9, Dewey's emphasis). Dewey writes in a similar sense that "The world is subject matter for knowledge, because mind has developed *in* that world; a body-mind, whose structures have developed according to the structures of the world in which it exists, will naturally find some of its structures to be concordant and congenial with nature, and some phases of nature with itself" (*EN* 211, Dewey's emphasis)—a sort of ecological, "esemplastic" mind.[18] Both Muir and Dewey find the distinction between *in* and *on* important enough to graphically emphasize *in*. We participate in the world.

Contrary to Gretel Ehrlich's comment in her introduction to the Penguin edition of *First Summer*, Muir experienced that first summer not as an "arrival" but as the onset of a long process that led forty years later to the person capable of writing *First Summer*. Muir and people in general participate *in* a process that for both Muir and Dewey is philosophical, aesthetic, and ecological. It is natural that our forms of knowledge and art emerge from a world already in process: "Art is thus prefigured in the very process of living. A bird builds its nest and a beaver its dam when internal organic pressures coöperate with external materials so that the former are fulfilled and the latter are transformed in a satisfying culmination. We hesitate to apply the word 'art,' since we doubt the presence of directive intent. But all deliberation, all conscious intent, grows out of things once performed organically through the interplay of natural energies" (*AE* 30). Dewey's language of typology and his organicism mark his own situatedness in a Romantic-Victorian tradition, but he too moves away from it while retaining a hold on the idea of organicism, still claiming that intent is linked to dynamic natural processes. Like his understanding of "unity," Dewey's organicism is a matter of process, cooperation, deliberation, and growth. It is an interplay of energy. It cannot be seen as static. Randall Roorda calls for a recuperation and a revision of organicism that both echoes and updates Dewey's claim above: "The need, in any case, is to divest organicism—the recognition of biological interconnectedness—of its hierarchical imperative," and he sees this need addressed in part by ecological writing (99). For both Muir and Dewey (and Roorda), art is inextricably linked to the changing, quotidian, natural world. The linkage of art and human existence with beavers and birds seems as radical here in Dewey as it often does in Muir's personifications of flowers and beasts. At the root of both conceptions is the leveling of art forms in an attempt to reinstate art in the everyday experience of people, and an extension of that experience so that it recognizes its interconnection and continuity with the physical world. All conscious intent, all forms of human expression, whether painting, poetry, philosophy, or nonfiction ecological writing, are embedded in that world so that all expression is a process of interaction between an agent and its environment: this is the art of knowing so essential to a pragmatist ecology.

 A carefully crafted textual moment is found in a passage from Muir's entry for 19 July. It epitomizes both Muir's writing and his mature relationship to the world. Beginning the entry with a description of the dawn's progressive colors, Muir proceeds to describe one of his favorite natural occurrences, a mountain storm,[19] drawing on firsthand experience to support his description: "the thunder gloriously impressive, keen, crashing, intensely concentrated, speaking with such tremendous energy it would seem that an entire mountain is being shattered at every stroke, but probably only a few trees are being shattered, many of which I have seen on my walks hereabouts strewing the ground" (*First Summer* 125). Through long experience, like Thoreau, who "was self-appointed inspector of snow storms and rain storms" (*Walden* 11–12), Muir establishes his expertise as an interpreter of the storm, and although he asserts its grandeur, he also is aware of its actual effects upon the woods. Then, "Now comes the rain, with corresponding extravagant grandeur, covering the ground high and low with a sheet of flowing water, a transparent film fitted like a skin upon the rugged anatomy of the landscape, making the rocks glitter and glow, gathering in the ravines, flooding the streams, and making them shout and boom in reply to the thunder" (*First Summer* 125). The rain flows over the surface of the landscape like a skin, its "rugged anatomy." Muir sees the landscape as a living thing in itself, capable of "glitter and glow," and its skin functions not as a boundary but as an interface between the human and the physical earth. We cannot be in the world without a body, and by endowing the mineral world with a skin and anatomy, Muir creates in this passage a world with a medium for having us, if you will, a world that seems to possess creative sentience.

The focus shifts from the water running over the ground to the water falling from the sky, and Muir begins a marvelous meditation on raindrops that is worth quoting here in full:

> Some, descending through the spires of the woods, sift spray through the shining needles, whispering peace and good cheer to each one of them. Some drops with happy aim glint on the sides of crystals, —quartz, hornblende, garnet, zircon, tourmaline, feldspar,—patter on grains of gold and heavy way-worn nuggets; some, with blunt plap-plap and low bass drumming, fall on the broad leaves of veratrum, saxifrage, cypripedium. Some happy drops fall straight into the cups of flowers, kissing the lips of lilies. How far they have to go, how many cups to fill, great and small, cells too small to be seen, cups holding half a drop as well as lake basins between the hills, each replenished with equal care, every drop in all the blessed throng a silvery newborn star with lake and river, garden and grove, valley and mountain, all that the landscape holds reflected in its crystal depths, God's messenger, angel of love sent on its way with majesty and pomp and display of power that make man's greatest show ridiculous. (126–27)

Muir pushes the limits of his rhetoric, but what at first seems like rhetorical excess is more like rhetorical plentitude. What holds the language in check is his insistence on alternating emotion with close observation and exact terminology: "happy aim" leads to "quartz, hornblende, garnet," and "blunt plap-plap" links up with "veratrum and saxifrage." By foregrounding both language and physical reality, Muir asserts an interrelationship by embedding language in the nonhuman world. The language is often rich and poetic, "sift spray through the shining needles," but it is still more than that. The raindrops, and Muir and his readers through them and their textualizaton, come to know the world with an intimacy beyond conventional human parameters, and it is this that causes joy and a sense of well-being. "Despite his occasional puritanical rants and his later environmental jeremiads, his words convey a joy, a playfulness that is lacking in most adult men, both then and now," writes John O'Grady (58). This joy emerges first from Muir's microscopic perception of rock crystals, and through this near view, the naming of the minerals becomes incantatory in itself. The raindrops know the cups of flowers, and kiss the lips of lilies, entering into an erotic, celebratory relation with the flora. They land in cells too small to be seen and in enormous lake basins, and the landscape itself binds this whole process together. Finally, the complete hydrological cycle is transformed by Muir's imagination into an enormous living entity, and, ultimately, it is the natural function that gives and sustains life and aesthetic experience. If and when it ceases to function, so does life on the planet.

Joy and beauty result from an ecological perception of the world. The hydrological cycle becomes a messenger of God, something far greater than "man's greatest show." Muir's words exist at the very heart of this text as an interface among the writer, the reader, the text, and the natural world upon which the first three depend. In these passages Muir refuses to succumb to the imperial imagination, and both fact and idea are brought to life in an ecological relationship. The basic fact—the hydrological cycle—is constelled with the idea—the angel of love—and both have equal value. The culture-based figure of the angel does not take precedence over the life-giving cycle of water. The nonhuman world and the world of the imagination create experience.

I suggested earlier that Muir's seemed to be an alpine imagination, and when Muir descends through the climate zones on the mountain his perception enfolds both the temporal and the spatial realms: "Descending four thousand feet in a few hours, we enter a new world—climate, plants, sounds, inhabitants, and scenery all new or changed" (*First Summer* 187). And finally:

> Looking up the cañon from the warm sunny edge of the Mono plain my morning ramble seems a dream, so great is the change in the vegetation and climate. The lilies on the bank of Moraine Lake are higher than my head, and the sunshine is hot enough for palms. Yet the snow round the arctic gardens at the summit of the pass is plainly visible, only about four miles away, and between lie specimen zones of all the principal climates of the globe. In little more than an hour one may swoop down from winter to summer, from an Arctic to a torrid region, through as great changes of climate as one would encounter in traveling from Labrador to Florida. (225)

The emphasis here on ecotonal boundaries is extraordinary, and Muir's achievement lies equally in his rendering of his own astonishment. The contrast of heat and snow activates the imagination, and the reader is placed in a position to "swoop" down like an eagle, transcending time—from winter to summer—and space—from Labrador to Florida. And this experience is not only literary—it is real, grounded in the process of vertical succession that makes the literary expression possible.[20] All manner of plants, animals, climates, and human inhabitants interact on this mountainside. Here in this alpine environment, the human being can experience the gradations of the nonhuman world, and in such a place one senses the rich potential of ecotonal relationships. On the western slope of the Sierra are "Hot deserts bounded by snow-laden mountains,—cinders and ashes scattered on glacier polished pavements,—frost and fire working together in the making of beauty" (228–29). We return to aesthetic process here, and as we have seen, Muir found the highest expression of aesthetics in natural processes—the hydrological cycle, the changing face of a rock, vertical succession—and here again beauty inheres in the interaction of natural elements made visible by the physical structure of the landforms and by the expression of the creative imagination. Aesthetic experience is forged in the ecological relationships among the human, nonhuman, and nonliving communities of the world. For Muir and Dewey, all beauty, all art, all human expression grows out of a physical world, sometimes just on the fringes of perception.

Muir's very first publication was the article "For the Boston Recorder.

THE CALYPSO BOREALIS. Botanical Enthusiasm. From Prof. J. D. Butler." It was taken from a letter Muir had written to his friend Jeanne Carr and sent by another friend, Professor James Davie Butler, to the *Boston Recorder*, which published it on 21 December 1866. Muir wrote the letter while in Canada dodging the draft during the Civil War, the utter breakdown of American democracy. War was to him an abomination. The letter details his discovery of the rare orchid *Calypso borealis*, or, ironically, "hider of the north," and Muir later claimed that the two great moments of his life were finding this flower and meeting Emerson in Yosemite (Wolfe, *Journals* 147). In the letter to Mrs. Carr, Muir's ecological ethos is already in process: "How good is our Heavenly Father in granting us such friends as these plant-creatures, filling us wherever we go with pleasure so deep, so pure, so endless. I cannot understand the nature of the curse, 'Thorns and thistles shall bring forth thee.' Is our world indeed worse for this 'thistly curse?' Are not all plants beautiful? Or in some way useful? Would not the world suffer by the banishment of a single weed?" (qtd. in Kimes and Kimes 1). Because they participate in experience, plants are afforded the ethical consideration normally reserved to humans, and Muir senses that "the curse must be within ourselves." The final question of the passage is one Muir put directly or implied in all his subsequent work. In the context of the war just ended in the United States, the question is poignant. If the world is impoverished by the loss of one weed, what can organized mass carnage mean for the world? What can the loss of human and nonhuman life caused by our politics and our ways of living in the world since Muir's time mean for the world of the future?

Muir leads us to possibility—a possibility of perception. He perceives the world in ecological terms and finds it joyous and beautiful. Frederick Turner writes that Muir "had reluctantly acted publicly on his love for the American land. His most significant discovery had been a way to love the land and to extend that love to the society at large" (340). We can see in Muir an extension of eros in Audre Lorde's sense of the term as "creative energy empowered" and the "sharing of joy" to the relationship between people and physical nature, and see in this extension a possibility to redress the original domination of the earth that seems to have underwritten our historical domination of women, peoples of color, and the impoverished (Lorde 4–5; see also Holmes 12, 244). To see ecologically is a goal, the consequences of which just might be the vision of communities living in a mutually beneficial way with other people and the natural world in a functioning democracy. Through his love and advocacy for the land, "In his own life he [Muir] had reconciled the conflict between democratic individualism and participatory democracy" (Turner 340). In cultivating ecological perception with its emphasis on interrelationships, we come to see that we are linked in a positive sense to *difference* itself. We are a long way from realizing this. But it is time to begin thinking and acting, as Barry Lopez says in *Arctic Dreams,* with a view of the future in which "we have the intelligence to grasp what is happening, the composure not to be intimidated by its complexity, and the courage to take steps that may bear no fruit in our lifetimes" (52). Dewey too urges us to embrace complexity. To help toward a productive perception of the world, he insists that "The only way to avoid a sharp separation between the mind which is the centre of the processes of experiencing and the natural world which is experienced is to acknowledge that all modes of experiencing are ways in which some genuine traits of nature come to manifest realization" (*EN* 30–31). In order to perceive the modern world in ways conducive to ongoing experience, we need both consciousness and the natural world intact—neither will function without the other—and we must realize that they are not completely separate entities. It is good to end here with a passage from Muir's journal that once again resonates in a strange way with Dewey: "There are no harsh, hard dividing lines in nature. Glaciers blend with the snow and the snow blends with the thin invisible breath of the sky. So there are no stiff, frigid, stony partition walls betwixt us and heaven. There are blendings as immeasurable and untraceable as the edges of melting clouds" (Wolfe, *Journals* 89).

John Steinbeck and Sparky Enea on the bridge of the *Western Flyer*, 1940. Courtesy of the Martha Heasley Cox Center for Steinbeck Studies, San Jose State University.

2
"The Form of the New"
Pragmatist Ecology and *Sea of Cortez*

> The striving of man for objects of imagination is a continuation of natural processes; it is something man has learned from the world in which he occurs, not something which he arbitrarily injects into that world.
>
> —John Dewey

> I was lost in the wonder of the extent to which all my life I have . . . unconsciously pragmatised.
>
> —Henry James

As John Dewey arrived at the height of his powers, John Steinbeck's life as a writer was in its infancy. In 1929 Steinbeck published *Cup of Gold,* followed by *The Pastures of Heaven* in 1932. *To a God Unknown,* his first work to voice an identifiable ecological outlook, appeared in 1933,[1] and an interesting but tenuous link can be drawn between it and Muir's *First Summer.* Along the North Fork of the Merced, Muir finds a rock in the stream that "is a nearly cubical mass of granite about eight feet high, plushed with mosses over the top and down the sides to ordinary high-water mark." He climbs on top of the rock, "with its mossy level top and smooth sides standing square and firm and solitary like an altar." Muir lies down on the rock, and "The place seemed holy, where one might hope to see God" (*First Summer* 48–49). One of the two central symbols in *To a God Unknown* is a big mossy rock in a glade with a spring running out from it, and "The green moss covering the rock was thick as fur, and the long ferns hung down over the little cavern in its side like a green curtain" (102). At the end of the novel, in order to save the drought-stricken land, Joseph Wayne climbs up and lies on the rock altar and sacrifices himself.[2] While an explicit connection is so far impossible to establish, it is nonetheless interesting to think that the sacrality of rock and water provides a common thread between two of California's great ecological writers.

Steinbeck's most famous work, *The Grapes of Wrath,* was published in 1939, and ecological devastation as much as anything else drives the Joads westward.[3] In his 1941 essay "The Philosophical Joads," Frederic Carpenter claims that Jim Casy was a pragmatist all along and that in Steinbeck's ideas "The mystical transcendentalism of Emerson reappears, and the earthy democracy of Whitman, and the pragmatic instrumentalism of William James and John Dewey" (242). Earthy democracy to be sure, but we need to clarify "instrumentalism." By "instrumentalism," Dewey never meant a narrow conception of the term as a way to create ideas to achieve narrow ends. Rather, for him it meant bringing intelligence to bear on the common events of everyday life and giving every citizen the tools to use his or her intelligence to best enhance meaningful experience—to use the art of knowing to live aesthetically.

Instrumentalism in its pragmatist context also has a democratic valence. In "The Development of American Pragmatism," Dewey argues that what pragmatists "insist upon above all else is that intelligence be regarded as the only source and sole guarantee of a desirable and happy future" ("Pragmatism" 19). Intelligence, as we have seen, is most active when it is engaged with an environment; for a pragmatist ecology, intelligence is useless when not engaged with both the physical and cultural environments. Dewey goes on to emphasize the interrelations of political, scientific, and aesthetic values in this pragmatist understanding of instrumentalism: "The more one appreciates the intrinsic esthetic, immediate value of thought and of science, the more one takes into account what intelligence itself adds to the joy and dignity of life, the more one should feel grieved at a situation in which the exercise of joy and reason are limited to a narrow, closed and technical social group and the more one should ask how it is possible to make all men participators in this inestimable wealth" (21). Art, science, and intelligence come together to activate the art of knowing. Instrumentalism does not use ideas to pave the way for a more efficient assembly line, for instance, but to enhance the value of life. As Hugh P. McDonald argues, in Dewey's thought, "Intrinsic values are not ultimate, but temporary consummations of experience, which are later instrumental to further consummations" (100). Value, then, is not located in an object but in a process. Again, McDonald can help: "Intrinsic value is naturalized by Dewey in his denial that it is fixed, transcendent, or outside natural processes" (108). Joy and dignity are key values, essential for the aesthetic consummation of human experience, which in its turn is rooted in the physical environment. Value is created through the art of knowing. Once again, Dewey claims that the art of knowing has been sequestered by a privileged class, devaluing experience for the many. He insists on a return of the art of knowing to the people and their everyday experience, just as Steinbeck insists on the dignity of the Joads and all the migrant farmers in *The Grapes of Wrath*.

Steinbeck seems to have been awake to the ideas and values of pragmatism. He was surely aware of William James's work; James had published *Pragmatism* in 1907, and as this chapter's second epigraph attests, pragmatism was very much in the air. Steinbeck read at least parts of *Pragmatism* as a student at Stanford (DeMott, *Steinbeck's Reading* 155), and *East of Eden*'s Samuel Hamilton not only owns James's *The Principles of Psychology* but claims that James is "a man the world is going to hear from" (188).[4] Dewey's *Experience and Nature* and *Art as Experience* were published at about the same time Steinbeck was hitting his stride as a novelist, and Steinbeck was reading Dewey's *A Common Faith* while he composed *East of Eden*. Clearly, Steinbeck was engaged by the American pragmatist tradition.

While he weathered the stormy reception of *The Grapes of Wrath* in 1939, including denunciation of the novel in Congress, Steinbeck became increasingly interested in marine biology and ecology, investing in Edward Ricketts's lab in Monterey, California.[5] He was also influenced, as Richard Astro shows, by Ricketts's favorite holistic thinkers: Jan C. Smuts, W. C. Allee, John Elof Boodin, and William E. Ritter (43–51).[6] But even before Steinbeck met Ricketts, he had certainly absorbed some of Ritter's concepts. In 1923 Steinbeck and his sister Mary took summer classes at Hopkins Marine Station near the Steinbeck cottage in Pacific Grove, and their marine zoology course was taught by C. V. Taylor, a PhD candidate in the Zoology Department at Berkeley, which was at the time deeply influenced by Ritter's thought (Benson 63). According to Brian Railsback, "Ritter's philosophy of the natural world emphasizes the interrelation of the whole of nature and all of its parts; the whole not only depends upon the parts from which it is made but also determines the direction of the parts" (14). Ritter's holism became a major component of Steinbeck's thinking. Boodin writes that "the laws of thought must be the laws of things" (xvi), and these men "are evolutionary thinkers who share John Dewey's feeling that the facts of evolution forced a modesty on philosophy by which it had acquired a sense of responsibility" (Astro 48–49). Darwin, as Jean Arnold points out, is an originary force in U.S. nature writing; he is conspicuously present in Dewey's philosophy as well. "In laying hands upon the sacred ark of absolute permanency," Dewey argues, "in treating the forms that had been regarded as types of fixity and perfection as originating and passing away, the *Origin of Species* introduced a mode of thinking that in the end was bound to transform the logic of knowledge, and hence the treatment of morals, politics and religion" ("Darwinism" 3). Science, morals, and politics exist in relation, though—and this is a crucial point—the driving force behind profound change is not scientific fact but a "mode of thinking." In the same essay, Dewey makes this point even more explicitly:

> Finally, the new logic introduces *responsibility* into the intellectual life. To idealize and rationalize the universe at large is after all a confession of inability to master the courses of things that specifically concern us. As long as mankind suffered from this impotency, it naturally shifted a burden of *responsibility* that it could not carry over to the more competent shoulders of the transcendent cause. But if insight into specific conditions of value and into specific consequences of ideas is possible, philosophy must in time become a method of locating and interpreting the more serious of the con-

flicts that occur in life, and a method of projecting ways for dealing with them: a method of moral and political diagnosis and prognosis. (13, emphasis added)

Along with a new evolutionary logic of physical existence evolves a new logic of responsibility for philosophy, for intellectual life, and for the general art of knowing. The art of knowing is not a matter of metaphysics—it is rooted in the world.

Darwin's presence permeates the text of *Sea of Cortez,* although it is most often the Darwin of *The Voyage of the Beagle* that so intrigued Steinbeck and Ricketts. Like Dewey, Steinbeck and Ricketts value less what Darwin discovered than how he discovered it, his "mode of thinking": "In a way, ours is the older method, somewhat like Darwin on the Beagle" (60). They reactivate the idea of Darwin as a naturalist, a term that was then falling from favor in the sciences, "and the modern process—that of looking quickly at the whole field and then diving down to a particular—was reversed by Darwin. Out of long, long, consideration of the parts he emerged with a sense of the whole" (60).[7] Darwin focused on the relations between species and their environments. As the facts of evolution do for philosophy, the facts of ecology inject a sense of ethical and moral responsibility and usefulness into *Sea of Cortez.*

Ritter and Darwin have an important bearing on the conception of *Sea of Cortez* in that, according to Ricketts, "inter-relation seems to be pretty much the key-note of modern holistic concepts, wherein the whole consists of the animal or the community in its environment" (qtd. in Hedgpeth 26). Interrelation is a central concept in *Sea of Cortez* and can only occur if boundaries are perceived as permeable, and the perception of interrelationships and porous boundaries becomes a crucial aspect of Steinbeck's aesthetics. Steinbeck and Ricketts write that "We are no better than the animals; in fact in a lot of ways we aren't as good" (69), and the sense of interrelationship among human and nonhuman communities so important to the aesthetics of *Sea of Cortez* carries over into an awareness of an ethical relationship: "It seemed to us that life in every form is incipiently everywhere waiting for a chance to take root and start reproducing; eggs, spores, seeds, bacilli—everywhere" (164). An ethical responsibility toward life avoids interfering unduly in its opportunity to evolve, and when humans invariably find themselves unable to fully understand the nonhuman world, "It is not enough to say that we cannot know or judge because all the information is not in" (165). Lack of final knowledge does not absolve one from taking action. One responds by posing further questions, and then one acts on the basis of available knowledge and the considered consequences of those actions: "An answer is invariably the parent of a great family of new questions" (166). In a pragmatist sense, unanswerable questions possess the greatest potential; they engage intelligence and enable human beings to change given situations. They enable the art of knowing to act on and with the world, and, of course, the world is always an unfinished thing. If the world were made complete, then no need for human intelligence would have arisen, no need for art, science, philosophy, or books about ecological writing. Steinbeck, Ricketts, and Dewey know that experience is open-ended. Whereas Dewey claims a moral and ethical use for philosophy, Steinbeck and Ricketts voice a similar call that includes philosophy, biology, and ecology.

In 1940, the same year Steinbeck received the Pulitzer Prize for *The Grapes of Wrath,* Steinbeck and Ricketts planned a biological collecting trip into the Gulf of California (the Sea of Cortez), of which *Sea of Cortez* became the published account. From 11 March to 20 April 1940, Steinbeck, Ricketts, Steinbeck's wife, Carol, and the master and crew of the *Western Flyer* sailed from Monterey to the Gulf of California, collected marine samples from the intertidal zones, and then returned to Monterey. In 1941, *Sea of Cortez,* written by Steinbeck and Ricketts, was published. In 1951 the narrative portion of the book, along with the memorial essay "About Ed Ricketts," appeared under only Steinbeck's name as *The Log from the Sea of Cortez.* In an early commentary on Steinbeck's work, Edmund Wilson observed that Steinbeck's art was driven by his interest in the biological aspect of the human race (42). Steinbeck, it seems safe to say, has managed, because of this focus, to alienate humanist critics such as Leslie Fiedler and Harold Bloom, who were unable to understand his interest in a biological dimension of human life and experience that he, like Muir, saw as not degrading to human beings but as the primary one linking humans to the rest of the physical world.[8] And for Dewey, not only the bodily portion of the human being was linked to the physical world, but also the reasoning part. In fact, as we have seen, he resists separating the two. In his 1938 book, *Logic: The Theory of Inquiry,* Dewey argues:

> It is obvious without argument that when men inquire they employ their eyes and ears, their hands and their brains. These organs, sensory, motor or central, are biological. Hence, although biological operations and structures are not sufficient conditions of inquiry,

they are necessary conditions. The fact that inquiry involves the use of biological factors is usually supposed to pose a special metaphysical or epistemological problem, that of the mind-body relation. When thus shunted off into a special domain, its import for logical theory is ignored. When, however, biological functions are recognized to be indispensable constituents of inquiry, logic does not need to get enmeshed in the intricacies of different theories regarding the relations of mind and body. (30)

On its most basic level, Dewey's contention is that separating mind and body is illogical. Our biology is evident in everything we do, and not only our physical selves but also our rational and creative selves are enmeshed in the biological world. Understood through the idea of pragmatist ecology, our science and our art participate in the ecologies around us. These biological and ecological principles contribute heavily to all of Steinbeck's work, but nowhere are they as important as in *Sea of Cortez*.

Two studies with far-reaching implications published in *Science* and *Nature* in 2001 can be usefully brought to bear on Steinbeck's biological perspective and on *Sea of Cortez*.[9] The studies sent reverberations through the scientific community by showing that human beings possess not around 142,634 genes, as previously hypothesized, but somewhere between 30,000 and 40,000, only half again as many more than a roundworm. Moreover, some parts of the human genome are inherited from bacteria and other external forms. This, of course, sets us much closer to the rest of the animal kingdom, just as Darwin did in 1859. The findings were, appropriately, announced on Darwin's birthday. Just as Darwin's did, these findings disturb many people who wish to retain rigid distinctions between the human and the nonhuman, who cannot bear the thought that humans are, simply put, animals. The discovery also breaks apart Cartesian, reductionist thinking, because human complexity cannot possibly be generated with so few genes in the ways science has traditionally explained our genetic makeup. To Stephen Jay Gould's mind:

> The collapse of the doctrine of one gene for one protein, and one direction of causal flow from basic codes to elaborate totality, marks the failure of reductionism for the complex system that we call biology—and for two major reasons. First, the key to complexity is not more genes, but more combinations and interactions generated by fewer units of code—and many of these interactions . . .

must be explained at the level of their appearance, for they cannot be predicted from the separate underlying parts alone. So organisms must be explained as organisms, and not as a summation of genes. Second, the unique contingencies of history, not the laws of physics, set many properties of complex biological systems. Our 30,000 genes make up only 1 percent or so of our total genome. The rest—including bacterial immigrants and other pieces that can replicate and move—originate more as accidents of history than as predictable necessities of physical laws. Moreover, these noncoding regions, disrespectfully called "junk DNA," also build a pool of potential for future use that, more than any other factor, may establish any lineage's capacity for further evolutionary increase in complexity.

Gould comments further that this "deflation of hubris is blessedly positive, not cynically disabling. The failure of reductionism doesn't mark the failure of science, but only the replacement of an ultimately unworkable set of assumptions by more appropriate styles of explanation that study complexity on its own level." This discovery underlines some basic ecological principles and helps to point up just how forward-looking Steinbeck and Ricketts's ecological thinking is in *Sea of Cortez*. Most importantly, our evolution is based on complexity and interaction, not on one-to-one correspondences. Organisms must be understood holistically, as organisms in environments, not as aggregates of parts, and this would seem to support Steinbeck and Ricketts's holistic outlook *and* the revision of hierarchical organicism that Roorda called for. This new information deflates hubris and its quixotic quest for certainty. Just as Dewey and James insist, contingency is the basic condition of human experience. And, our basic genetic structure has a significant component that comes not from us but from the external physical world; the evolution of our bodies, just like that of our culture, is sustained by "immigrants"—those who cross permeable boundaries.

Along with this stance, Steinbeck and Ricketts call for a widening of scientific interpretation that can extend Gould's comments about combinations and interactions in genetics:

For example: the Mexican sierra has "XVII-15-IX" spines in the dorsal fin. These can easily be counted. But if the sierra strikes hard on the line so that our hands are burned, if the fish sounds and nearly escapes and finally comes in over the rail, his colors pulsing and his tail beating the air, a whole new relational externality has come into being—an entity which is more than the sum of the fish plus the fisherman. The only way to count the spines of the sierra unaffected by this second relational reality is to sit in a laboratory, open an evil-smelling jar, remove a stiff colorless fish from formalin solution, count the spines, and write the truth, "D. XVII-15-IX." There you have recorded a reality which cannot be assailed—probably the least important reality concerning either the fish or yourself. (*Cortez* 2)

The prose technique infuses the scientific endeavor with a conversational tone that brings the scientist and the famous novelist onto the same page with the everyday reader, perhaps an angler. There are multiple, relational realities, and the one that includes a complexity involving fish, boat, life, death, color, and physical sensation is privileged over a narrow, reductionist conception of scientific truth, which, for Steinbeck and Ricketts, and for an ecological understanding of experience, is of minimal importance. Observer and observed, subject and object are necessarily part of the same process. The fish in the jar of formaldehyde has a history and a life that Steinbeck and Ricketts seek to recover, and in this effort they blur the boundaries between scientific writing and common experience. In a comment that Rachel Carson will echo in the next chapter, Steinbeck and Ricketts write: "It has seemed sometimes that the little men in scientific work assumed the awe-fullness of a priesthood to hide their deficiencies, as the witch-doctor does with his stilts and high masks, as the priesthoods of all cults have, with secret or unfamiliar languages and symbols. It is usually found that only the little stuffy men object to what is called 'popularization,' by which they mean writing with a clarity understandable to one not familiar with the tricks and codes of the cult. We have not known a single great scientist who could not discourse freely and interestingly with a child" (*Cortez* 73). *Sea of Cortez* resists scientific hubris, and within the book the spaces between human and animal, boat and sea, physical nature and human expression, text and world, become spaces charged with evolutionary potential for a renewed ecological perception of the natural environment and pragmatism's role in that reconsideration. The writer, or the scientist, or each of us, "brings his own limitations to the world. If he has strength and energy of mind the tide pool stretches both ways, digs back to electrons and leaps space into the universe and fights out of the moment into non-conceptual time. Then ecology has a synonym which is ALL" (85). This inclusiveness seems also a democratic value, an integral part of a pragmatist ecology emphasizing the permeability of mind, body, and universe. In order to embrace in written form an ecology so defined, Steinbeck and Ricketts decided to loosen up the form of the book, to let the book form itself, "its boundaries a boat and a sea" (1), and as soon as the *Western Flyer* reached the open sea, "The forward guy-wire of our mast began to sing under the wind, a deep and yet penetrating tone like the lowest string of an incredible bull-fiddle" (30). The simile turns the machine into a musical instrument, into a tool for both scientific and aesthetic understanding. Steinbeck and Ricketts's language allows an image of the trip as art to resonate through the rest of the book and beyond—across subjectivities and cognitive frameworks.

I have talked about the concept of ecotones—the border between a ship and a bass fiddle, perhaps—as highly productive places both aesthetically and ecologically, and suggested that at an ecotone, aesthetics and ecologies can interrelate. *Sea of Cortez*, the material text, can also be seen as textually ecotonal. It is a document in which various forms, styles, and references interpenetrate. This book of nearly six hundred pages begins with the narrative account of the collecting trip, the one later published under Steinbeck's name as *The Log from the Sea of Cortez*. The narrative is followed by a short "Note on Preparing Specimens" that is followed by eight color plates and thirty black-and-white plates, picturing many specimens of the more than 550 species collected on the trip. Then come an extensively annotated phyletic catalog, a glossary, and an index. Ecological concepts permeate the narrative, which in its turn enhances the beauty of the pictures and catalogs of the second section of the book. This book is, I think, a unique artifact. It puts a different spin on the genre of nature writing. It pushes nature writing toward ecological writing. The fruit of a collaboration between one of the most famous novelists of the twentieth century and an accomplished marine biologist, it is in many ways a precursor of our attempts to examine how science and literature can inform each other. There is a permeable boundary between the words of the scientist and those of the novelist. In a memorandum to his editor, Pat Covici, Steinbeck emphasizes just that point: "Originally a journal of the trip was to have been kept by both of us, but this record was found to be a natural expression of only one of us. This journal was subsequently used by the other chiefly as a reminder of what actually had taken place, but in several cases parts of the original field notes were incorporated into the final narrative, and in one case a large section was lifted verbatim from other published work. This was then passed back to the other for comment, completion of certain chiefly technical details, and corrections. And then the correction was passed back again" (qtd. in Astro 13). Although the manuscript pages of the narrative portion are in Steinbeck's hand,[10] both the form and the content of the book are informed by ecological ideas, so in that spirit I refer to both Steinbeck and Ricketts as the authors. The book's form is ecological—the form of the new.

As a work of nonfiction, *Sea of Cortez* has been deeply mined for clues to Steinbeck's philosophy and worldview as they bear on his fiction. It is, however, essential to "recognize that *Sea of Cortez* was conceived not as a supplement to Steinbeck's fiction, but as a work of feeling, meaning, and beauty in its own right" (Perez 47). Donald Culross Peattie writes in the *Saturday Review of Literature* that "The grandest of passages in this strange bargain-buy of two full-length books under one cover . . . are some of the descriptions of the tide, mirages, wind or dead calm, Indians, tidal pools, or simply night, or dawn, or the barking of a dog, or loneliness on a hot and empty shore. Style, reality, *Stimmung* at these moments, all deepen and intone" (7).[11] Peattie is correct, of course, except his assessment misses the crucial point that *Sea of Cortez* is not two books but one, and that the scientific matter contributes to the intonation of all the moments. Without the science, the moments would not exist in their profundity, a conclusion Stanley Brodwin seems to support: "the poetry of the genre is created by more than a sense of epic adventure; it resides more profoundly in 'scientific thinking,' the methodology of the scientific mind that controls the purposes of the narrator struggling to penetrate, if not resolve, questions of still greater mystery" (147). Steinbeck and Ricketts know that although "the process of gathering knowledge does not lead to knowing" (*Cortez* 165), action directed by thought does, the bringing to bear of intelligence on the world. In *Sea of Cortez*, the process of asking questions and experiencing intensely a particular ecosystem leads to the perception of the mystery of a 1941 "world picture not dominated by Hitler and Moscow but something more vital and surviving than either."[12] It seems that the idea of ecology undermines authoritarianism. The mystery of the littoral zones in *Sea of Cortez* includes the relationships between a phyletic catalog, a dog's bark, and a lonely church in a dusty town. The links among these things inform the methodology of *Sea of Cortez,* and to a large degree, Steinbeck, freshly grounded in marine biology following the tumultuous reception of *The Grapes of Wrath,* articulates the beauty of these components through his perception of ecological relationships, both natural and textual.

As early as 1939, Steinbeck was seeking a new form of expression to follow his great novel. He writes to Carlton Sheffield that "I've worked the novel—I know it as far as I can take it. I never did think much of it—a clumsy vehicle at best. And I don't know the form of the new but I know there is a new which will be adequate and shaped by the new thinking."[13] The extension of ecological principles of interrelationship to scientific writing seems the major component of "the form of the new," and the "new thinking" likely refers to the holistic thought of Boodin, Smuts, and others. However, Steinbeck is also clearly indebted to a native philosophical tradition, leading to the pragmatism of James and Dewey. He would have felt quite at home in pragmatism's insistence on the all-important interrelationships of science, philosophy, and literature.

Of particular interest in relation to *Sea of Cortez,* Dewey insists upon the central importance of experience understood as perception of relations between creature and environment undergoing dynamic change, a process that "has pattern and structure, because it is not just doing and undergoing in alternation, but consists of them in relationship" (*AE* 50–51). Dewey denies that matter belongs to the mind, insisting that "That to which both mind and matter belong is the complex of events that constitute nature" and arguing that philosophical quarrels "go on within the limits of a too domestic circle, and can be settled only by venturing further afield, and out of doors" (*EN* 66, 47). Like Dewey, Steinbeck and Ricketts in *Sea of Cortez*—certainly an "out of doors" book—refute philosophical systems too narrowly formulated, seeing them primarily as markers of intellectual timidity: "when the horizons stretch out and your philosopher is likely to fall off the world like a Dark Ages mariner, he can save himself by establishing a taboo-box. . . . Into this box he can throw all those thoughts which frighten him and thus be safe from them" (*Cortez* 54). When Steinbeck in his letter to Sheffield mentions form, he means a fluid, permeable form—a "complex of events." For both Steinbeck and Dewey, form is not a box; form flows and shifts, just as do, for example, the relationships within the human genome, between communities within a particular ecosystem, and between ecosystems themselves.

After collecting around San Francisquito Bay, Steinbeck and Ricketts sense a shifty formal relationship between language and natural world, and merge text with species: "Our own interest lay in relationships of animal to animal. If one observes in this relational sense, it seems apparent that species are only commas in a sentence, that each species is at once the point and base of a pyramid, that all life is relational to the point where an Einsteinian relativity seems to emerge. And then not only the meaning but the feeling about species grows misty. One merges into another, groups melt into ecological groups until the time when what we know as life meets and enters what we think of as non-life: barnacle and rock, rock and earth, earth and tree, tree and rain and air" (216). Just as Muir saw everything "hitched to everything else in the universe," their focus is on relationships, not on individuals in isolation. In the introduction to the phyletic catalog, Ricketts concurs: "our purpose was, primarily, to get an understanding of the region as a whole, and to achieve a toto-picture of the animals in relation to it and to each other, rather than to amass a great collection of specimens" (304). Their concern is ecological, and their language in particular enhances a perception of ecological relations. Notably, they align species with language, with the punctuation marks that ease the perception of relations between words and syntactical structure. Species are the commas that allow one to perceive whole ecosystems. They then metaphorically include geometry in the formula with the reference to the pyramid but also seem to insist upon the possibility of something being at once part of the base and the point of the pyramid. This paradoxical statement, which, like a comma, gives us pause, provides also a moment to sense relationships. Along with Einstein, physics and relativity enter into the discussion, and meaning and feeling begin to meld. One cannot be divorced from the other; emotion cannot be driven from scientific investigation. There is no pure scientific objectivity ("The limitation of the seeing point in time, as well as in space, is a warping lens" [264]), and no theorist can look in from without, removed from the fact of his or her own looking.

Finally, Steinbeck and Ricketts choose words that suggest an interchange between feeling and meaning, between life and nonlife, in terms that are sacro-sexual: "one merges into another," "groups melt into," "life meets and enters what we think of as non-life." The phrase "what we think of" casts doubt on the idea that anything is nonliving. Only through theoretical positioning do we term entities like rain and air and soil nonliving when they are exactly the elements that sustain life. They may be as much a part of our own living organism as are testicles and ovaries. With the metaphor of species as commas, and with comments such as "Indeed, as one watches the little animals, definite words describing them are likely to grow hazy and less definite" (207), Steinbeck and Ricketts acknowledge that we are trapped to a large degree within language, but language here forms a mediational node of experience shared among reader, writer, and physical world. The perception of these relations contributes to ways of thinking, to a pragmatist ecology that contributes to the attempt to discover ways to live in the world that honor the essential relations among human and nonhuman entities. Human language is necessarily part of any ecosystem we engage, and the language of interrelationship found in *Sea of Cortez* makes a contribution to efforts to know the physical and cultural world as a field of permeable boundaries, as a complex of events where all forms of life, not simply the human, evolve.

The art of knowing is articulated in this language of interrelationship. Interrelationships arise within nature, and nature is an ongoing process—it is contingent. To deny the contingency inherent in any way of knowing is to deny that knowing exists within the context of the natural world. "Only if the one who engages in knowing be outside of nature and behold it from some external locus can it be denied that knowing is an act which modifies what previously existed, and that its worth consists in the consequences of the modification," Dewey claims in *The Quest for Certainty* (195). However, as Dewey well knew, the observer or knower cannot exist outside the physical world. Knowing does not depend upon some objective knower discerning an a priori truth. There is no such thing; all knowing proceeds in an environment, "all meanings intrinsically have reference to natural events" (*EN* 219). Meaning grows from relationships, particularly from the relationship between a creature and its environment, and the expression of knowing and meaning creates change. In a book where Steinbeck and Ricketts take us "into the Sea of Cortez, realizing that we become forever a part of it; that our rubber boots slogging through a flat of eel-grass, that the rocks we turn over in a tide pool, make us truly and permanently a factor in the ecology of the region" (*Cortez* 3), we begin to understand that environment, creature, and the creature's art are all entwined in a pragmatist ecology of aesthetic experience. In much the same way that Michael Branch sees John Muir linked to contemporary ecology, Steinbeck and Ricketts anticipate recent thinking that nestles the human in the physical environment: "contemporary ecological theory assumes that the individual is not an autonomous locus of power but is rather an expression of nature fundamentally inextricable from its ecosystemic context" (Branch, "Telling" 100).

For Dewey, Muir, Steinbeck, Ricketts, and for a pragmatist ecological definition of experience, these webs of connections between the individual and other living and nonliving "expressions of nature" are centrally important. Experience has an aesthetic value contingent upon the quality of the relations sustained between the self and the environment. There is a pushing back and forth, a tension, between self and world that must be maintained. This interface of self and world is the source of art ecologically considered, of the art of knowing. The relations between the creative intelligence and the physical world are ongoing, and "In the degree . . . in which the mind is weaned from partisan and egocentric interest, acknowledgement of nature as a scene of incessant beginnings and endings, presents itself as the source of philosophical enlightenment" (*EN* 83). In an enlightened aesthetic experience, Dewey writes, "The *material* out of which a work of art is composed belongs to the common world rather than to the self, and yet there is self-expression in art because the self assimilates that material in a distinctive way to reissue it into the public world in a form that builds a new object" (*AE* 112, Dewey's emphasis), the form of a book perhaps. If the assimilated material of a book such as *Sea of Cortez* is the material of ecosystemic interrelationships—a form of enlightenment that roots knowledge itself in the relations between the human creature and its environment—then once the book enters the public realm, the aesthetic representation of ecological values has an ecosystemic function within the community. The book becomes a thing that is both the culmination and the beginning of a natural history.

The notions of interface and ecosystemic function must be extended here to include not only the self and the physical world but also the public world or community. Community, physical environment, and self, then, are inextricably linked. One of the first laws of ecology states that we cannot do only one thing. As we know from Edward Lorenz and subsequent chaos theory, because many things are connected, when we disturb the most insignificant-seeming member of an ecosystem or a culture, the ripple effect of our meddling can be so far reaching and complex that the potential damage caused is beyond our limited knowledge of how the world functions. As we help speed the physical world's deterioration, on a global or local scale, a degraded common world inevitably results. Deterioration and degradation of the environment that is the ultimate source of all experience are certain to negatively affect the quality of our experience as both public and individual human beings in ways as yet impossible to foresee. An impoverished experience will yield damaged communities and diminished art. Art and community are lodged in the physical world, and, for Dewey and for us, the link between the stewardship of the natural world and democracy then becomes explicit: "Conservation of not only the public domain but restoration of worn-out land to fertility, the combating of floods and erosion which have reduced vast portions of our national heritage to something like a desert, are the penalties we have to pay for past indulgence in an orgy of so-called economic liberty. Without abundant store of natural resources, equal liberty for all is out of the question. Only those already in possession will enjoy it. Not merely a modification but a reversal of our traditional policies of waste and destruction is necessary if genuine freedom of opportunity is to be achieved" ("Freedom" 251). Once again, Dewey argues for a way of thinking and being in the world that benefits the majority of the people, and that way depends upon a physical environment that is intact—that is nurtured in common. The role of the individual is not to be at liberty to engage in promiscuous self-enrichment but rather to be free to be the strongest possible contributor to the community at large. There is no self outside of community and environment.

The aesthetic quality of experience and its art can mark an ecologically sound relationship to the world, and a contingent and collective notion of truth expressed by philosophical pragmatism can help articulate a community's relation to the world. David Abram has argued that "Ecologically considered, it is not primarily our verbal statements that are 'true' or 'false,' but rather the kind of relations that we sustain with the rest of nature. A human community that lives in a mutually beneficial relation with the surrounding earth is a community, we might say, that lives in truth" (264).[14] Relations are central to experience, so we can say that truth is seated in the experience of community. "Common" and "community" share the same root, *communis,* which, according to the *OED,* indicates a "community of relations or feelings" along with a sense of obligation. Dewey's common world re-merged with Abram's community may give rise to art that enhances the truth of a democratic community. Ecological writing, at least in part, is an expression of this merger, and among its possible consequences is a community more fully experiencing its environing world. And that nearby world is essential to democratic communities: "Democracy must begin at home, and its home is the neighborly community" (*PP* 368). Steinbeck and Ricketts sense the pattern of this communal experience at the end of *Sea of Cortez:* "We had all felt the pattern of the Gulf, and we and the Gulf had established another pattern which was a new thing composed of it and us" (267–68). A community evolves among the people aboard the *Western Flyer,* the Gulf itself, and, I would add, the text of *Sea of Cortez*. "A new thing," a pragmatist ecological community, evolves when consideration of community participation is extended to the physical world itself.

Ecological writing is surely the expression of an individual author's experience of the natural community, the culmination of a natural history. But as Jack Turner comments, that experience is of value to a community only when shared: "Yet most of us, when we think about it, realize that after our own direct experience of nature, what has contributed most to our love of wild places, animals, plants—and even, perhaps, to our love of wild nature, our sense of our citizenship—is the art, literature, myth, and lore of nature. For here is the language we so desperately lack, the medium necessary for vision" (89). The individual's relation to the natural world does not develop in isolation. Dewey and Turner clearly understand that experience is both in itself aesthetic and enhanced by productions of the common culture, in this case the art of knowing, or environmental art. All art is rooted in an environment. Dewey has this to say on the matter: "It is a commonplace that we cannot direct, save accidentally, the growth and flowering of plants, however lovely and enjoyed, without understanding their causal conditions. It should be just as commonplace that esthetic understanding—as distinct from sheer personal enjoyment—must start with the soil, air, and light out of which things esthetically admirable arise. And these conditions are the conditions and factors that make an ordinary experience complete" (*AE* 18). Art is inseparable from its natural, environing world. Art and experience have an ineradicable connection to the physical world, and environmental art foregrounds this connection. Dewey continues: "The first great consideration is that life goes on in an environment; not merely *in* it but because of it, through interaction with it. No creature lives merely under its skin; its subcutaneous organs are means of connection with what lies beyond its bodily frame, and to which, in order to live, it must adjust itself" (19, Dewey's emphasis). This interpenetration of individual and environment is the condition for the survival of community, democracy, and art. The devaluation of the environing world, then, degrades what we supposedly hold most dear: democracy, communities, and creative intelligence.

Dewey goes on: "The career and destiny of a living being are bound up with its interchanges with its environment, not externally, but in the most intimate way" (*AE* 19). Neither Steinbeck nor Ricketts nor Dewey limits the term "environment" to the physical world, and Steinbeck and Ricketts much admire a small, intimate church in the little town of Loreto with a plaster Madonna in a side chapel. The statue performs an integral function in the region's ecology: "This lady, of plaster and wood and paint, is one of the strong ecological factors of the town of Loreto, and not to know her and her strength is to fail to know Loreto. One could not ignore a granite monolith in the path of the waves. Such a rock, breaking the rushing waters, would have an effect on animal distribution radiating in circles like a dropped stone in a pool. So has this plaster Lady a powerful effect on the deep black water of the human spirit" (175). Human manipulation of plaster, wood, and paint is aligned with a natural, monolithic landform, and the effect of the statue on the town is aligned with the effect of wave shock on animal populations. Spirituality and marine biology converge in the local ecology: the cultural and physical worlds in which we live permeate our most intimate core and are thus necessarily part of our quotidian existence and part of every utterance and act. Art is integral to any ecology in which humans participate. As the scientist pulling a Mexican sierra from a jar of formaldehyde uncovers an unassailable, useless truth, when we separate culture from ecology, we too discover at best a partial truth of limited use. Truth should be understood as a kind of action directed toward community participation that sustains a mutually beneficial relationship to cultural and physical environments. Anything that enhances such a relationship contributes to the health of the community. In this sense, ecological writing and environmental art—and all art in a pragmatist ecology is environmental—can perform the cultural work of community building. To borrow words from Turner, the art of nature affords us the "vision" to see and articulate "citizenship" in the natural community.

Growing critical and creative interest in environmental art seems to parallel the need that many of us feel for a reconstruction of our communities. Particularly relevant to *Sea of Cortez*, this interest also functions to revitalize texts that are seen to function in the way Turner suggests, or, as Barry Lopez writes in *Arctic Dreams*, texts that help us see the land with "a deeper understanding of its nature, as if it were, itself, another sort of civilization we had to reach some agreement with" (12). Notably, both Lopez, perhaps our preeminent contemporary environmental writer, and Utah writer Terry Tempest Williams have cited Steinbeck as a seminal influence on their work. In his collection of autobiographical essays, *About This Life,* Lopez writes about spending time at summer camp with Steinbeck's sons and goes on to say, "Before I left for college, I read all of Steinbeck's books. I drew from them a sense of security" (9). In "The Outside Canon," Williams writes that "Steinbeck gives us clues as to how a healing can occur—and in fact it demands nothing less than our blood" (71).[15] Obviously, Steinbeck has had strong influence among the ranks of U.S. ecological writers.

In "What Is to Be Done with the Biosphere?" David Raines Wallace makes an intriguing observation that locates *Sea of Cortez* in a pivotal space among ecological texts: "The Second World War marked a fundamental change in the human condition. Before it, the human relationship with other humans was the crucial ethical and philosophical concern. After it, the human relationship with the living Earth, the biosphere, became the crucial concern" (94). Steinbeck had the uncanny ability (or luck) to publish *The Grapes of Wrath* at the precise moment when the historical conditions that made the novel possible were poised to disappear. Likewise, *Sea of Cortez,* though rather unluckily, was published the first week of December 1941, the same week the Japanese bombed Pearl Harbor, precipitating U.S. entrance into World War II. Steinbeck again lands right on the cusp of a changing world. He is aware of the dramatically heightened mechanistic exploitation of the natural world, yet the threat of nuclear devastation does not yet exist. Steinbeck and Ricketts can still write, "We have made our mark on the world, but we have really done nothing that the trees and creeping plants, ice and erosion, cannot remove in a fairly short time" (*Cortez* 88). Of course, we now know that this no longer obtains, and along with our knowledge evolves our increasing awareness of the importance of an ecological reconceptualization of communities and art, of a pragmatist ecology.

Although Steinbeck and Ricketts seem to believe, along with Emerson, that our "operations taken together" can only insignificantly and temporarily damage the natural world, they begin to sense the consequences stemming from contemporary abuse of the environment. The crew of the *Western Flyer* encounters a large fleet of Japanese fishing vessels, complete with factory ship, and gain permission to go aboard one of the vessels. They watch the dredge come up, and "The sea bottom must have been scraped completely clean. The moment the net dropped open and spilled this mass of living things on the deck, the crew of Japanese went to work. Fish were thrown overboard immediately, and only the shrimps kept" (248). Appalled at the eradication of all life on the sea floor in collusion with the Mexican authorities, Steinbeck and Ricketts comment: "We liked the people on this boat very much. They were good men, but they were caught in a large destructive machine, good men doing a bad thing. With their many and large boats, with their industry and efficiency, but most of all with their intense energy, these Japanese will obviously clean out the shrimps of the region. And it is not true that a species thus attacked comes back. The disturbed balance often gives a new species ascendancy and destroys forever the old relationship" (249). Steinbeck and Ricketts become here, in contrast to their earlier statement, aware that conventional wisdom concerning human inability to permanently damage the biosphere is untenable. About our own culture, they point out that "With our own resources we have been prodigal, and our country will not soon lose the scars of our grasping stupidity" (250). Soon, even the Monterey sardine fishery that supported the *Western Flyer* and its crew will be fished out. Steinbeck and Ricketts's maintenance of the shrimpers' innocence is quickly undone by knowledge that neither their guilt nor their innocence bears relevance to a problem whose consequences extend far beyond the culpability of individuals. They realize that the shrimpers "were committing a true crime against nature and against the immediate welfare of Mexico and the eventual welfare of the whole human species" (250). Their crime is against the relations between humans and their physical environment.

In Deweyan terms, the relationship between the shrimp boats and the world is one that fouls potential experience, hence not a true relationship. Experience is dependent upon a creature's interaction with its environment, and the degradation of that environment disallows the evolutionary potential of countless creatures that may have contributed to an ongoing experience. Experience's inability to complete itself leads to unbalance, which "blurs the perception of relations and leaves the experience partial and distorted, with scant or false meaning" (*AE* 51). In the grim efficiency of the shrimp fleet's exploitation of the sea for economic gain, critical ecosystems that both sustain the lives of sea animals and provide a valuable food source for natives of the region are destroyed. The ecological aesthetic of balanced interaction between creature and world is devastated.

By this balance I mean a shifting balance ever at play and requiring continual adjustment to changing conditions. Dewey's aesthetics have an ecological valence, "since," as he says, "man succeeds only as he adapts his behavior to the order of nature" (*AE* 154) and not the other way around. "Underneath, the rhythm of every art and of every work of art there lies, as a substratum in the depths of the subconscious, the basic pattern of the relations of the live creature to his environment" (155). Dewey's use of the terms "adapts," "rhythm," "pattern," and "relations" clearly indicates a fluctuating balance that requires continual adjustment of our being to allow play and interpenetration between person and world. This adjustment allows an extension of ethical consideration to the natural world as a participant. Steinbeck and Ricketts often imply this stance when they blur distinctions and taxonomies. Our ways of thinking about the nonhuman world are too narrow: "as species merges into species, the whole idea of definite independent species begins to waver, and a scale-like concept of animal variations comes to take its place. The whole taxonomic method in biology is clumsy and unwieldy, shot through with the jokes of naturalists and the egos of men who wished to have animals named after them" (*Cortez* 207). Again, as one looks more closely, humanly constructed categories begin to blur and merge. Extended, this passage merges the human observer with the animals he or she observes.

On the other hand, imposing our taxonomies and our egoism (and our sense of humor) on the nonhuman world seems to diminish it. Granted, some measure of this is unavoidable, and when all is said and done, environmental discourse in the United States could benefit from a more healthy sense of humor. One of the major (perhaps the most important) results of Steinbeck and Ricketts's trip into the Gulf was "no service to science, no naming of unknown animals, but rather—we simply liked it. We liked it very much" (*Cortez* 270). Although one reviewer who liked the book found its jokes worn out—"The boating magazines long ago extracted all the risibles from outboard motors and the *Coast Pilot,* but as the authors are new to the sea this can be forgiven them" (Lyman 183)— the pleasure involved in the experience is critical to its understanding. That distinctions tend to merge upon close involvement with the nonhuman is, in *Sea of Cortez,* a pleasurable experience. It is not something to be frightened of, not something to scare humans into destructively protecting their identity, an identity that is often predicated upon assumed human superiority to other living and nonliving things. Steinbeck and Ricketts see that these things "were all one thing and we were that one thing too" (270). Their pleasure results from the realization that human beings, right down to the bacteria in our genetic pool, are part of a much larger world. If we are indeed parts of a larger thing, then it follows that it behooves us to extend ethical consideration to all participants and parties of the larger community. This seems like the pleasure of ecological democracy. We might like it.

Lawrence Buell argues for a mode of expression that "opens itself up as well as it can to the perception of the environment as an actual independent party entitled to consideration for its own sake" (*Ecological Imagination* 77). Buell's language confers political status on the environment, and Steinbeck and Ricketts do indeed acknowledge nature as plaintiff, an entity with rights, just as the nation of Mexico and the human race have rights. The Japanese shrimp boats deprive the local ecosystem of its right to experience full evolutionary potential, hence the crime. All experience is an ongoing process, culminating only by enabling other new experiences. The experience is ethically sound if it enhances our relationship to the world and at the same time maintains the potentiality of shifting relationships and continued experience. Experience is ethical when it seeks a productive ecological balance, though necessarily temporary, that is interpenetrative, interactive. Experience approaches the greatest good only when it is shared.

Because Dewey's aesthetics embraces what seems an environmental ethics, his aesthetics, as Thomas Alexander suggests in another context, is not about art, but about human life in relation to the environing world (*Horizon* 60). This environing world is, for Dewey, "the whole complex of the results of the interaction of man, with his memories and hopes, understanding and desire, with that world to which one-sided philosophy confines 'nature.' The true antithesis of nature is not art but arbitrary conceit, fantasy and stereotyped convention" (*AE* 156). Dewey's statement places reductionist, Cartesian notions about the duality of nature and humans, of body and mind, under erasure, freeing up space for an aesthetics underpinned by ecological principles. At the opening of *Art as Experience*, Dewey links geographers, geologists, philosophers, and art (9–10), and the ecological representation of human beings in their environing world, as Dewey's linkage clearly implies, insists upon integration of the methods of science and literature; Steinbeck "found a great poetry in scientific writing" (Fensch 31), and he worked this poetry by rendering journals, scientific texts, historical works, the *Coast Pilot,* and the captain's log into the textual ecotones of *Sea of Cortez.*

In fact, Steinbeck and crew go on a six-week voyage with a four-hundred-pound box of books. Two of these books have been little remarked upon:

the U.S. Hydrographic Office's *Coast Pilot: H.O. No. 84, Sailing Directions for the West Coasts of Mexico and Central America from the United States to Colombia Including the Gulfs of California and Panama* (1937) and the Jesuit priest Don Francisco Javier Clavigero's 1786 treatise *The History of Lower California*. Both are mentioned from the onset of chapter 1 in *Sea of Cortez*, and Steinbeck and Ricketts refer to the *Coast Pilot* at least sixteen times and to Clavigero at least twelve. Both texts are critical parts of the ecology of the book.

These two books, upon closer inspection, like Steinbeck and Ricketts's species, sometimes tend to merge. The *Coast Pilot* is a series of official guides for navigation in coastal waters. While it functions as an actual navigational tool in the Gulf of California, and while it helped Steinbeck and Ricketts re-navigate the Gulf when they began composing *Sea of Cortez*, its attempts at pure objectivity inevitably fail. Clavigero's *History* provides an added spiritual dimension that Steinbeck and Ricketts integrate into their ecological text. Of course, the *Coast Pilot* guides the ship and crew from point to point, from Monterey to San Diego, around Cape San Lucas, up the western shore of the Gulf from La Paz to Angel de la Guardia, and down the eastern shore to Guaymas, Agrabampo Estuary, Espiritu Santo Island, and back out to sea and home. The anonymous authors of the *Coast Pilot* have little tolerance for variance from the factual, and "The *Coast Pilot* spoke as heatedly as it ever does about mirage and treachery of light" (*Cortez* 6). The *Coast Pilot* despises that which cannot be defined, but some skeletal descriptive prose manages to creep in nevertheless. For instance, "The coast from Cabo Falso to Cape San Lucas, a distance of about 4 miles, is a succession of sand beaches and *bold* rocky cliffs against which the sea breaks *heavily*," and "Cape San Lucas is a headland of *fantastic* shape, formed by two high, *bold* rocks of *grotesque* appearance" (108, emphasis added). The prose is often flowing and the reading quite interesting, coastlines are minutely detailed, and Steinbeck and Ricketts's descriptions jibe for the most part with the *Coast Pilot*.

The *Coast Pilot*, like the ship, is an indispensable tool, but it functions beyond its primary use as a guide to navigation. It mentions customs, hours, pilot's fees, and health and communication services ashore. Steinbeck and the crew of the *Western Flyer* follow malaria up and down the coast of Lower California and Sonora, and Steinbeck and Ricketts write, "we picked up the malaria on the other side, ran it down to Topolobambo, and left it there. We would say offhand, never having been to either place, that the malaria is very bad at Mulege and Topolobambo"

(184). The 1937 *Coast Pilot* says nothing about malaria at Mulege, but it says this about Topolobambo: "no effort is made at sanitation. The vicinity is infested with gnats and mosquitoes, and malarial fevers are prevalent" (196). It sets rather clear boundaries—it tells the crew where to go or not go, when, and how much it will cost. Even though Steinbeck and Ricketts complain about the rigidity of the *Coast Pilot*, it clearly contributes to their perception of the ecologies of both the Sea of Cortez and the *Sea of Cortez*.

However, Steinbeck and Ricketts are more drawn to Clavigero: "Going back from the *Coast Pilot* to Clavigero, we found more visual warnings in his accounts of ships broken up and scattered, of wrecks and wayward currents; of fifty miles of sea more dreaded than any other" (6). Clavigero's lively narrative of soldiers and Jesuit missionaries on the peninsula and in the surrounding sea at times better advises Steinbeck and Ricketts than does the *Coast Pilot*. The interpenetration of these two texts is not unlike the permeable nature of different sections of *Sea of Cortez*. Steinbeck and Ricketts set their margins with "the *Coast Pilot*, like an elderly scientist, cautious and restrained, on one side—and the old monk, setting down ships and men lost, and starvation on the inhospitable coasts" on the other (6). They situate themselves, the expedition, and the reader between these two sources—an elderly scientist and an elderly priest—in a space (like the chapel in Loreto) where spirit and science tend to share a common ecological vision: "And it is a strange thing that most of the feeling we call religious, most of the mystical outcrying which is one of the most prized and used and desired reactions of our species, is really the understanding and the attempt to say that man is related to the whole thing, related inextricably to all reality, known and knowable" (216–17). Here the space between the *Coast Pilot* and Clavigero, between and within texts, expands into infinity, into the "ALL," into the realm of possibility, and echoes with the intuition of the inextricable connections gearing human beings onto the natural world and its creatures.

Ecofeminist critic Freya Matthews writes that the world "appears as a field of relations, a web of interconnections, which does indeed cohere as a whole, but within which a genuine form of individuation is nevertheless possible. An individual . . . maintains itself by way of its continuous interaction with its environment" (239), and we hear echoes of Dewey, Muir, Steinbeck, and Ricketts in her words. The interaction with environment is an ecotonal process, and it seems clear that Steinbeck and Ricketts's understanding of ecology evolves at least in part from the ecotone where science and spirit intersect. Therefore, Steinbeck and Ricketts's ecological poetics could not risk dualistic, positivist science. Positivist science at its worst tends toward a narrow, supposedly objective view of the world, and as I have been arguing, Steinbeck, Ricketts, and Dewey mean to break the world out of systems narrowly conceived, to break the sierra out of its jar of formaldehyde. As Peter Englert claims, *Sea of Cortez* "could not possibly be the work of narrow academic minds" (191), because its inclusion, along with great prose, of everything from philosophy and science to raucous parties creates a unique aesthetic experience in which scientific and literary discourse reciprocate. The interaction of scientific and literary form with the natural world provides, then, a wider view of a "complex of events"—it activates the art of knowing.

The risky aspect of scientific method alone is that it becomes enshrined, that the "complex of events" becomes frozen. In an article examining metaphor in scientific and literary discourse, Liliane Papin writes: "In quantum physics . . . the concept of event has emerged to replace the concept of thingness" (1255), the verbal sense of process thus supplanting the substantive meaning of a specific state of being. Similarly, "There is one great difficulty with a good hypothesis," write Steinbeck and Ricketts. "When it is completed and rounded, the corners smooth and the content cohesive and coherent, it is likely to become a thing itself, a work of art. It is then like a finished sonnet or a painting completed. One hates to disturb it" (*Cortez* 180). A hypothesis, in other words, is only effective and beautiful if it is capable of continual modification, if it functions more as a verb and less as a noun, if it participates in ongoing experience. In a pragmatist ecology, art is also understood as a process, not an object.

Dewey, too, advocates scientific method as a way of thinking, not as a dogma; he concerns himself with its working, not with its thingness, with science as a verb, not a noun. The world is an ongoing process, and the continued formulation of hypotheses, tests, and conclusions is primarily an application of critical intelligence. So too for Steinbeck and Ricketts, who seem to have redefined art conceived of as a finished thing. *Sea of Cortez* is a form of the new: "We wanted to see everything our eyes would accommodate, to think what we could, and, out of our seeing and thinking, to build some kind of structure in modeled imitation of the observed reality. We knew that what we would see and record and construct would be warped, as all knowledge patterns are warped" (2). Shifting, bending patterns, not disciplines, define knowledge, and Steinbeck, Ricketts, and

Dewey erect methods or structures suited to an environment in which all things, all truths, all art, all knowledge are contingent.

All three men search not for transcendent ideals but for truth or belief that can work for us, which is to say, truth or belief let out of its box. Rather than looking for first causes, a pragmatist inquiry critically examines the consequences of a particular concept. William Chaloupka points out that "no experience worth having . . . could be spoken of as coming from nowhere or having no consequence past its own duration" (251). That an idea or text contributes to ongoing experience questions the fixed nature of boundaries and borders. Physical borders and textual boundaries permeable as the human epidermis, as the boundaries of animals that do not distinguish themselves from their territories, are locations of heightened potentiality.[16] In an essay about literature of the American West, Stephen Tatum envisions such boundaries as "topographies of transition" whose borders are "membranes" through which seemingly oppositional elements interact (325). While similar concepts can be found, for instance, in Bakhtin's boundary-inhabiting word and Homi Bhabha's transformation of national boundaries into "internal liminality,"[17] Tatum levels the ontological, epistemological, and ecological and refuses to grant privilege to any one category. In this view all boundaries are ecotonal—fringes, places of enhanced evolutionary potential for live creatures, habitats, texts, sensations, knowledge systems, and beliefs.

The littoral zone is just such an ecotonal space between the highest and lowest reaches of the tides. It is, then, a confluent ecotone where the temporal and the spatial literally flow together. For Steinbeck, Ricketts, and their readers it is where most of the activity in *Sea of Cortez* takes place; it shifts with every wave, and every minute the relentlessness of the sea alters the physical characteristics of the place as it transports sand, leaves tide pools, exposes rocks and sea life, then covers them again. The Sea of Cortez is itself a place of mirage and shifting coastlines, "The maps of the region were self-possessed and confident about headlands, coastlines, and depth, but at the edge of the Coast they became apologetic—laid in lagoons with dotted lines, supposed and presumed their boundaries" (*Cortez* 6), and as the *Coast Pilot* warns, bars "and channels are undoubtedly subject to change" (191). The dotted line—the dash—signifies a permeable boundary; *Sea of Cortez* assimilates the material of the world, and in it are moments and places where thought, physical nature, community, and art enrich one another in a fecund ecotonal matrix.

In a textual moment of cooking and eating as communal effort, we revisit Dewey's notion that the material of art is the common world, not the self, although the self assimilates that material in a distinctive way. With the ship at anchor in Puerto San Carlos, the crew of the *Western Flyer* sees the water filled with schools of voracious fish, feeding upon each other, and "Sparky went to the galley and put the biggest frying pan on the fire and poured olive oil into it. When the pan was very hot he began catching the tiny fish with the dipnets, a hundred or so in each net. We passed the nets through the galley window and Sparky dumped them in the frying pan. In a short time these tiny fish were crisp and brown. We drained, salted and ate them without any cleaning at all and they were delicious" (*Cortez* 239–40). The nets, themselves matrices of dotted and knotted lines, are permeable objects, tools for taking nourishment from the sea, able to pass easily through the water's surface. The fish are then passed through the open boundary of the galley window, into the hot oil encompassed by the pan, and then into the human body through the open mouth. The fish taken from the sea pass through liminal, permeable surfaces preceding their transformation into energy and their eventual return to the sea as food for micro-organisms. We can link this cycle of nourishment to the form of *Sea of Cortez*, "its boundaries a boat and a sea" (1), both boundaries, as we see here, quite permeable.

This nourishing crossing of boundaries and surfaces, passing through planes of water, through the technological spaces of ship's window and frying pan, through the surface of olive oil, and through the mouth into the human body and out, is reiterated and extended. After the meal, Steinbeck and Ricketts see the schools of fish as "A smoothly working larger animal surviving within itself—larval shrimp to little fish to larger fish to giant fish—one operating mechanism. And perhaps *this* unit of survival may key into the larger animal which is the life of all the sea, and this into the larger of the world" (241, emphasis in source). The ongoing existence of the world and the text depends upon the permeability of forms, and the human ingestion of fish is aligned with the ecological relationship between species of sea life, then extended to embrace the world in a profound realization of the interconnectedness of all things, of the astonishing and reaffirming "knowledge that all things are one thing and that one thing is all things—plankton, a shimmering phosphorescence on the sea and the spinning planets and an expanding universe, all bound together by the elastic string of time" (217). Human participation in the confluence of time, space, matter—in a complex of events, in experience.

Steinbeck and Ricketts realize that "It is advisable to look from the tide pool to the stars and then back to the tide pool again" (*Cortez* 217).[18] The movable head and eyes of the human being connect sky and sea, stars and earth, and the human body—interface of skin, orifice, pore, and cornea—is infinitely permeable. Dewey concurs: "The epidermis is only in the most superficial way an indication of where an organism ends and its environment begins" (*AE* 64). Experience completes itself "through what the environment—and it alone—can supply, . . . [through] a dynamic acknowledgment of this dependence of the self for wholeness upon its surroundings" (65). Although we no longer argue for a whole self, the sense of Dewey's statement is clear: the surfaces of the self and the world interpenetrate, "There are things inside the body that are foreign to it, and there are things outside of it that belong to it" (64). All experience depends upon this most common, but most commonly denied, interdependence of self, community, world, and even the bacteria in our genetic structure.

Steinbeck and Ricketts did more than unconsciously pragmatize. William James's appearance as recommended reading in *East of Eden* and Steinbeck's nomination of Dewey as one of those who provides "a goal for others to shoot at, as an enticement to effort for the world's good" (qtd. in DeMott, *Steinbeck's Reading* 141) further seem to indicate that some circulation of pragmatism, and especially of Dewey's thought, is more prevalent in Steinbeck and Ricketts's ecological aesthetics of experience than we have assumed. At the close of the narrative, Steinbeck and Ricketts again transform their voyage into art: when the *Western Flyer* reaches the open sea on its homeward-bound leg, they echo their opening simile, and "the big guy-wire, from bow to mast, took up its vibration like the low pipe on a tremendous organ. It sang its deep note into the wind" (271), and the cycle of the trip is a musical cycle, the cycle of human aesthetic expression, the life cycle of the littoral zones, the "form of the new." Steinbeck and Ricketts write that "We at least have kept our vulgar sense of wonder" (69), and the art of *Sea of Cortez* is the art of life, of lived, shared human experience, and of wonder at the joyful participation of a writer, a scientist, a boat, a text, and a reader in the ecologies of the physical world.

Rachel Carson at a microscope, 1951. Photograph by Brooks Studio. Used by permission. Photo courtesy of Yale Collection of American literature, Beinecke Rare Book and Manuscript Library.

3
Rachel Carson's Marginal World
Pragmatist Ecology, Aesthetics, and Ethics

> The land ethic simply enlarges the boundaries of the community to include soils, waters, plants, and animals, or collectively: the land.... In short, a land ethic changes the role of Homo Sapiens from conqueror of the land-community to plain member and citizen of it. It implies respect for his fellow-members, and also respect for the community as such.
> —Aldo Leopold

> We lie, as Emerson said, in the lap of an immense intelligence. But that intelligence is dormant and its communications are broken, inarticulate and faint until it possesses the local community as its medium.
> —John Dewey

In 1941, the same year Steinbeck and Ricketts published *Sea of Cortez,* a marine biologist with the U.S. Fish and Wildlife Service, thirty-four-year-old Rachel Carson, published her first book, *Under the Sea Wind.* In this chapter, I had planned to discuss only *Under the Sea Wind* because it so closely relates in time and subject to *Sea of Cortez*—so closely, in fact, that *Under the Sea Wind* and *Sea of Cortez* were reviewed together in the *Saturday Review of Literature* in December 1941. William Beebe wrote of Carson's book, "the purpose of this book is to make the sea and its life a vivid reality, says Miss Carson, and she has succeeded" (5). It became increasingly difficult and, finally, impossible to isolate one of Carson's three books about the sea: *Under the Sea Wind, The Sea around Us* (1951), and *The Edge of the Sea* (1955). The books themselves are parts of an ecological literary project, and it seems counter both to Carson's spirit and to the ethos of this study not to engage all three books as interrelated, as constituting among them an ecology. The three books about the sea trace a process not unlike Muir's writing late in life his history of that first year. Like Muir, Carson develops her ecological aesthetics and ethics over time. *Under the Sea Wind* is her first major step in building an ecological aesthetics, predicated on the perception of interrelationships, that underpins the environmental ethic for which her best-known book, *Silent Spring,* is so famous. Although Carson's development through these books has been discussed before, I depart from previous critical work by enlisting Dewey's pragmatist aesthetics to help better understand Carson's work as building a pragmatist ecology that remains a powerful tool toward the creation of a feasible environmental ethic for our time.[1] Such an ethic embraces a community as Leopold describes it and links creative intelligence to local communities, both natural and cultural, as Dewey insists. In a reading of Carson's work inflected with Dewey's aesthetics, one comes to understand the vitality and the central importance of a breakdown of distinctions between subject and object, individual and community, human and nonhuman, science and lay knowledge that contributes to a pragmatist ecological understanding of experience that integrates science, aesthetics, and ethics with the everyday world.

I look at Carson's progression from *Under the Sea Wind* through her penultimate book, *The Edge of the Sea.* The emphasis of the discussion is brought to bear on *The Edge of the Sea,* a book that, like *Sea of Cortez,* attends primarily to the littoral zones, this time those of the Atlantic coast. *Silent Spring* receives only brief mention here, although it remains the book immediately associated with Carson's name. For instance, in 1999 the *New York Times* published a list of the top-hundred nonfiction books of the twentieth century, and *Silent Spring* ranks fifth behind such notable works as *The Education of Henry Adams*, *The Varieties of Religious Experience*, *Up from Slavery*, and *A Room of One's Own* ("Another Top 100"). Although such lists need to be taken with a grain of salt, it is nonetheless important that *Silent Spring* finds itself among books that are concerned with the dawning of a technological age, with a pragmatist philosopher's thoughts on religion, and with both the rights of an oppressed minority and the rights of women. *Silent Spring* certainly takes technology to task, there is a strong pragmatist bent to it, and it is chiefly concerned with ending the oppression of the living entity that supports all life, the physical world. It is certainly arguable that oppression of all kinds begins with the human being's attempt to dominate nature itself.[2] *Silent Spring* has, in fact, been called the *Uncle Tom's Cabin* of the environmental movement.[3] But, I am more interested at this stage in how Carson arrived at a position from which *Silent Spring* was possible. What can be learned from the way pragmatist ecology evolves from an aesthetic appreciation of the natural world?

For one thing, like the aesthetics of Steinbeck and Ricketts, Carson's aesthetics embraces scientific knowledge, and we see developing in the nature essay—and the ethical structure it often unfolds—the need to integrate science into aesthetic representations of the physical world. This, of course, enriches ecological writing, and it also has a deep impact on the public perception of science. As Vera Norwood points out in *Made from This Earth,* "Making completely new knowledge of genetics and chemistry available to laypeople, Carson helped deflate the worship of science prevalent in post–World War II America" (169). Also, Carson understood that without a clear sense of the observer's role in a work of ecological writing, it is impossible to construct an ethic that includes both the human being and the natural world. Her work always attempts a pragmatist ecology that keys into the relationships between the creative imagination and the natural and cultural environments. Thomas Lyon finds this focusing outward from the strictly human the great heresy of nature writing, and hope for an ecologically sustainable world likely lies precisely in this heresy (19). It is a heresy, of course, because it undermines our anthropocentric conception of the world. Nevertheless, Carson possessed a strong faith in science, yet in the end she found it morally and ethically impossible not to confront that faith when she felt compelled to challenge the government- and industry-sponsored mainstream scientific establishment of her day in *Silent Spring*.[4]

It is worth remembering that *Silent Spring*, and Carson's work in total, has a renewed relevance at the opening of the twenty-first century. Carson's pragmatist ecology rings with ethical ramifications. When *Silent Spring* appeared in 1962, Carson, a trained biologist, called to account the U.S. Department of Agriculture, science in the service of large industry, and powerful chemical producers on the issue of pesticide use in the United States. She finally brought about a shift in public opinion and widespread reform in the use and marketing of pesticides in the United States. DDT use was banned in the United States in 1972.[5] Her primary heresy was questioning the scientific, industrial, bureaucratic—and primarily male—authorities of her time, most often on their own terms. With this in mind, environmental historian Donald Worster writes in *Nature's Economy* that "the scientific conscience she symbolized became the central creed of the environmental movement: a vision of the unity of life, as taught by science, and a moral ideal of living cooperatively with all members of the natural community" (24).[6] This conscience is as important now as it was then. However, it needs to be dragged out of the symbolic realm and put into use. For example, some of the same companies producing pesticides when *Silent Spring* was published are now producing genetically modified foods and seeds.

Although *Silent Spring* is her best-known book, Carson was a well-established nature writer long before she began to write her book on pesticides. Unfortunately, her earlier work still tends to be overshadowed. In fact, her first publication, "A Battle in the Clouds," appeared in a children's magazine when she was ten years old.[7] From an early age she possessed both a love of the natural world and a love of writing and reading, both legacies from her mother, for whom Carson cared until she was nearly ninety years old. Writing was for her a lifelong process, and although in 1925 she entered Pennsylvania College for Women (now Chatham College) to study English, she changed her focus to biology, finally earning a master's degree in marine zoology from Johns Hopkins University in 1932. She later wrote, "I had given up writing forever, I thought. It never occurred to me that I was merely getting something to write about" ("Real World" 149).[8] I hear in this comment echoes of John Muir, who famously claimed, "I only went out for a walk, and finally concluded to stay out till sundown, for going out, I found, was really going in" (Wolfe, *Journals* 439). Both statements suggest that attention outward toward the natural world, beyond the purely human, is also a way inward. This looking outward makes their aesthetics and their public commitment to the natural world possible, and their questioning of anthropocentrism seems, again, the great heresy of ecological writing and of environmentalism in general.

When Carson graduated from Johns Hopkins in 1932, she entered the job market as a woman scientist in the thick of the depression. During this difficult time she supported both her parents. Then her father died in 1935, and the following year her sister also died, leaving two daughters whom Carson and her mother agreed to raise. After piecing together a living, she finally landed a job with the Bureau of Fisheries in 1935 as a technical writer and often wrote scripts for bureau radio broadcasts. One of the very few women hired by the bureau in a professional capacity, she worked for the federal government for sixteen years, eventually rising to editor in chief of Fish and Wildlife Service publications. In fact, her first literary success grew from her government work.

Carson originally wrote "Undersea" as the introduction to a Bureau of Fisheries publication, but her supervisor, Elmer Higgins, declared the work too literary for a government publication. He suggested that Carson submit it to the *Atlantic Monthly*, which she did, and it was published in the September 1937 issue. Carson later said that from "Undersea," "everything else followed." "Undersea" took up only four pages of the *Atlantic Monthly*, but most of Carson's concerns are clearly evident in this early piece. Primary among them is her ability to present complex scientific information to a general reading public, indeed to render science into poetry. Much of her writing about the sea attains the status of an extended prose poem.

"Undersea" opens with a rhetorical question: "Who has known the ocean?" Of course, no human being can know the ocean as its inhabitants do, and Carson employs in this piece the strategy she hones in her first book, *Under the Sea Wind:* the attempt to narrate from an underwater perspective. "Undersea" is a poetic submarine grand tour in which Carson evokes the strangeness of ocean life: "Dropping downward a scant hundred feet to the white sand beneath, an undersea traveler would discover a land where the noonday sun is swathed in twilight blues and purples, and where the blackness of midnight is eerily aglow with the cold phosphorescence of living things" (7). While Carson's prose brings the undersea world vividly before the reader, she is also careful to metaphorically link the sea world to the more familiar land world of her readers. The ocean is likened to pastures, the shifting tides to night and day. Although she insists that "to sense this world of waters known to the creatures of the sea we must shed our human perceptions of length and breadth and time and place" (4), the verbal connections to familiar phenomena such as "night," "day," "pasture," and tides that "abandon pursuit" and "fall back" establish a common bond between her readers' quotidian experience and a habitat so strange as to be beyond the imagination. Further, she conveys scientific material in a way that sparks the imagination: "the sea performs a vital alchemy that utilizes the sterile chemical elements dissolved in the water and welds them with the torch of sunlight into the stuff of life" (6). Carson's prose itself performs a "vital alchemy" by making available to a wide reading public the wonder of the sea, "the slow swells of mid-ocean" (4), and its processes, "one by one, brilliant-hued flowers blossom in the shallow water as tube worms extend cautious tentacles" (6).

Carson's prose is exact, careful, and lyrical. Without overtly calling attention to the science of ecology, she nonetheless constructs a picture of a world in which "Individual elements are lost to view, only to reappear again and again in different incarnations in a kind of material immortality" (11). This is the great cycle of life that so fascinated Carson and which she so loved. This unmitigated love of the world and its creatures—from fish to insects to human beings, although "chief, perhaps, among the plunderers is man, probing the soft mud flats and dipping his nets into the shallow waters" (6)—drives all of her work. Coupled to that love, Carson's commitment to communicate complex ideas to the nonspecialist and her attempt to make the material of science available to the public at large through its rendering into evocative prose remains consistent throughout her career. Such continuity reveals a deeply rooted dedication to ecology and to a democratic community. Through the worst times of personal pain and public attack, Carson remained unshaken in her belief that the human and the nonhuman worlds participate in experience and that both warrant ethical consideration because of that participatory role. Before her time, perhaps, she strove to see beyond the human.

Picking up where "Undersea" left off, Carson's first book, *Under the Sea Wind,* peers out beyond the human. It appeared just about a month before the attack on Pearl Harbor, and though well reviewed, *Under the Sea Wind* did not sell. It is deceptive in its simplicity. Fred D. White finds it the most compelling of Carson's early books: "The use of narrative and fable instead of the more customary exposition and exemplification for scientific discourse serves to fulfill Carson's twofold goal of (1) restimulating the sense of wonder that permits a spiritual/esthetic link with nature, and (2) disengaging nature from our 'meddling intellect' so that nature may once again be perceived as a transcendent realm, infinitely beyond our comprehension, not to mention our control" (190).[9] White makes secure claims for physical nature as something beyond our understanding, and, certainly, we could use a more spiritual, aesthetic link with the natural world. We need to re-evoke a sense of wonder in us all.[10] However, I do not think Carson constructs a narrative or fable about nature.[11] Nor is it helpful to think of human intellect, especially Carson's, as "meddling." The intellect is our most important tool: it activates the art of knowing; it engages the creation, ensuring the continuity of experience. Nor does "transcendent" necessarily mean something "infinitely beyond our comprehension." In Dewey's terms, the transcendent is the juncture—the fringe—just outside of sight that occasions the perception of relations. According to Victor Kestenbaum, "Dewey's pragmatism, viewed not simply as a philosophical theory but as an encompassing view of the human condition, is a generalization of the idea that our lives are lived in the realms and intersections of the tangible and the intangible," especially, to my mind, in the intersections (*Grace* 33). Carson, if she sought the transcendent at all, understood it as the juncture between the tangible and the intangible, in *Under the Sea Wind* as the juncture between the world of the creative human intelligence and the world of sea creatures. In this sense, then, the transcendent can be redefined as the perception of an ecological zone of and in transition. A work of ecological writing, as Scott Slovic argues, reveals not only the author's close attention to natural features but also the author's process of coming to awareness in the company of the natural world, of her or his negotiation of the space "between aesthetic celebration and scientific explanation" (4). Certainly, Carson employs personification and narrative in order to provoke awareness in her audience. But she moves beyond these strategies. Narrative is important; it is among our primary strategies for knowing our world. But there are other worlds. In her later work, Carson often finds these other worlds of the sea and shore rhapsodic precisely because they do *not* possess a narrative, precisely because of their *otherness:* "In the intertidal zone, this minuscule world of the sand grains is also the world of inconceivably minute beings, which swim through the liquid film around a grain of sand as fish would swim through the ocean covering the sphere of the earth" (*Edge* 130). The simile does not link the animal to the human but rather the animal to the animal, and unlike Blake's world in a grain of sand, the world Carson perceives is profoundly nonhuman. She has no desire to split aesthetics and science; following Kestenbaum, perhaps the transcendent is in the blurry transitional zone between art and science. Carson is after all both scientist and creative writer, and through both celebration and explanation she participates in the experience of the littoral.

The three sections of *Under the Sea Wind* narrate the life cycles of shore birds, of mackerel, and of eels. According to Linda Lear, Carson claimed that "taken together, the three narratives would weave a tapestry in which the ecology of the ocean and the interdependence of all its creatures would emerge" (90).[12] This is not a linear narrative. It is an interdependence of narratives, and the narratives attempt to voice the intersection of the tangible and the intangible. Carson's choice of a tapestry as a metaphor for a text is a traditional one; however, it is brought up to date by the introduction of ecology into the threads of her text. In weaving her narratives, Carson is careful to avoid claiming human motivation, consciousness, and emotions for her nonhuman characters. She does give her fish and bird characters names, and she struggles to see from their perspective. She does, of course, attempt to evoke an emotional response to these creatures from her audience in an effort to establish an interdependence of reader, text, and sea life. Rather than sacrifice the otherness of the creatures in her book, Carson attempts to create a node of mediation between writer, the reader, and the world of the ocean creatures, an intersection between the creative intelligence and a participating world. In short, her work contributes to ongoing experience.

In the foreword to the 1941 edition of *Under the Sea Wind*, omitted in reissues of the book, Carson writes: "I have spoken of a fish 'fearing' his enemies . . . not because I suppose a fish experiences fear in the same way that we do, but because I think he *behaves as though he were frightened*. With the fish, the response is primarily physical; with us, primarily psychological. Yet if the behavior of the fish is to be understandable to us, we must describe it in the words that most properly belong to human psychological states" (qtd. in Lear 91).[13] To a large degree, then, we cannot avoid textualizing. She uses literary strategies to help her readers identify with her nonhuman characters and remains accurate to the scientific facts available to her at the time.[14] Chapter 7, "The Birth of a Mackerel," is a good example. In it Carson writes: "The next three days of life brought startling transformations. As the processes of development forged onward, the mouth and gill structures were completed and the finlets sprouting from back and sides and underparts grew and found strength and certainty of movement. The eyes became deep blue with pigment, and now it may be that they sent to the tiny brain the first messages of things seen" (*Sea Wind* 127). Carson selects accurate details of the physical development of a fish from a hatchling, and the sprouting of fins and gills is decidedly other-than-human. But she then sharpens the focus on the deep blue eyes of the baby mackerel, zooming in on the most humanlike characteristic of the profoundly other-than-human fish and leaving the reader with the suggestion that this tiny creature has an awakening perception distinctly its own, a perception beyond human knowledge but no less important. Perhaps because of the subtle link Carson makes through her description of the fish's eyes, we find it at least plausible that the mackerel has some sort of consciousness. Yet, it remains most fascinating precisely because it is not like us, so nonhuman. We find ourselves participating in the juncture of knowable and unknowable; again, in Dewey's sense—and for a pragmatist ecology—this is as far as the idea of transcendence goes, to the ecotone of the tangible and the intangible. At this node, relations are perceived and experience takes flight from its perch.

Human beings appear in *Under the Sea Wind* infrequently, usually taking on the role of predators. The book also has a minimum of authorial intrusions, and when they do occur they function to emphasize the importance of the animals' own context. For example, Carson has this to say about fish migrating from the sea back into the rivers: "By the younger shad the river was only dimly remembered, if by the word 'memory' we may call the heightened response of the senses as the delicate gills and the sensitive lateral lines perceived the lessening saltiness of the water and the changing rhythms and vibrations of the inshore waters" (17). She again suggests an alignment of human memory and perception with the physical sense organs of the fish. Their world is one dominated by instinct, but just the same, that instinct is valued on a similar plane to human memory and perception. In this narrative of instinct and interconnections, often with fish in the subject position, our universal privileging of human structures gives way to an aesthetics that values the intersections of human knowledge and the life of the sea.

The sea is an astonishing interface of ecosystems: "There could be scarcely a stranger place in the world in which to begin life than this universe of sky and water, peopled by strange creatures and governed by wind and sun and ocean currents. It was a place of silence, except when the wind went whispering or blustering over the vast sheet of water, or when sea gulls came down the wind with their high, wild mewing, or when whales broke the surface, expelled the long-held breath, and rolled again into the sea" (*Sea Wind* 117). This is a world wholly governed by air, water, and light and inhabited by birds, fish, and mammals; it is a vast sheet from which human influence is absent or minimal, although the verbs "peopled" and "governed" function to create a verbal link between the human sphere and the "stranger" spaces of the nonhuman. Its silence encourages attentiveness, and the ecotonal surface of the water serves as a metaphor for the interaction of wind and water, of flying and swimming creatures who, in breaking the surface of the sea, signify interdependence. Carson's art, like that of Steinbeck and Ricketts, helps us to experience the sea and its life-forms in the way that Dewey claims art functions: "Through art, meanings of objects that are otherwise dumb, inchoate, restricted, and resisted are clarified and concentrated, and not by thought working laboriously upon them, nor by escape into a world of mere sense, but by creation of a new experience" (*AE* 138). This happens in *Under the Sea Wind* partly through a presentation of nonhuman subjectivity, and Carson's woven narratives about sea creatures introduce her readers to an experience of the sea itself. Here, the sea becomes a universe—an experiential space—teeming with ecological and aesthetic meaning. For Carson, at least in 1941, these ecotonal spaces of the sea—brimming with evolutionary and creative potential—seem impervious to lasting human damage.

When Carson began to think about her next book, she envisioned a more comprehensive work about the sea itself, still clinging to the belief that humans "cannot control or change the ocean as, in his brief tenancy of earth, he has subdued and plundered the continents" (*Sea around Us* 18). Although prematurely optimistic, *The Sea around Us* is an attempt to grasp the whole of the sea, and Carson is more inclusive than she was in her previous book. *The Sea around Us* is also more distanced in point of view than *Under the Sea Wind*. Carson used thousands of sources during her research for the book, and *The Sea around Us* both includes the most up-to-date science of her time and alludes to the literature and mythology of the sea. In the preface to the 1961 edition of *The Sea around Us*, Carson warns of the increased dumping of radioactive materials into the sea prompted by Cold War research: "It is a curious situation that the sea, from which life first arose, should now be threatened by the activities of one form of that life. But the sea, though changed in a sinister way, will continue to exist; the threat is rather to life itself" (xiii).[15] Moving beyond *Under the Sea Wind*, the pragmatist ecology of *The Sea around Us* involves not only the landforms and inhabitants of the oceans but also human perceptions, representations, and behaviors in relation to the marine environment. In fact, we have more in common with the creatures of the sea than we care to admit: "each of us begins his individual life in a miniature ocean within his mother's womb, and in the stages of his embryonic development repeats the steps by which his race evolved, from gill-breathing inhabitants of a water world to creatures able to live on land" (14). Whereas in *Under the Sea Wind* Carson sought primarily an emotional connection, from *The Sea around Us* on, like Steinbeck and Ricketts, she insists also on a biological one.

For Carson, then, human expression is part of the ecology of the sea, which she develops further in her praise of the *Coast Pilot* throughout *The Sea around Us*. Carson's use of the *Coast Pilot* also establishes another interesting link to Steinbeck and Ricketts in *Sea of Cortez* and to their perception of the juncture between scientific description and aesthetic representation.[16] Like Steinbeck and Ricketts, Carson finds the *Coast Pilot* an indispensable practical tool and a rich source of material and evocative prose. In fact, at the end of *The Sea around Us* she traces the genealogy of the *Coast Pilot* series, and she quotes liberally from them: "Besides giving detailed accounts of the coastlines and coastal waters of the world, these books are repositories of fascinating information on icebergs and sea ice, storms, and fog at sea. Some approach the character of regional geographies" (246). Carson even compares *Coast Pilot: Sailing Directions for the Northwest and North Coasts of Norway* to Poe's description of the Maelstrom (158). Steinbeck, Ricketts, and Carson had a sense that texts such as the *Coast Pilot* could provide information not only about coastlines but also about how the most rigorous attempts at scientific objectivity must integrate the most unpredictable phenomena. For instance, sometimes animals can help in the largely technical task of navigation: "These flocks and the direction of their flight on approaching, together with the use of the lead, are of great value in making the island when it is foggy," and "Blowing whales usually travel in the direction of open water" (qtd. in *Sea around Us* 211).

It is precisely when the authors of the *Coast Pilot* are at their most unsure that Steinbeck and Ricketts find them most interesting: "We trust these men. They are controlled, and only now and then do their nerves break and a cry of pain escape." They cry out because "No matter how hard they work, the restlessness of nature and the carelessness of man are always two jumps ahead of them" (*Cortez* 107). Science cannot fully contain the local, variable landscape, which, though frustrating to scientists like the authors of the *Coast Pilot,* fascinates Carson, Steinbeck, and Ricketts. Their aesthetics inhere in the intersection of the tangible and the intangible. Carson too isolates points where the *Coast Pilot* authors resort to seeking local knowledge because "In phrases like these we get the feel of the unknown and the mysterious that never quite separates itself from the sea" (*Sea around Us* 211). In other words, the meaning of a particular place overflows the limits of scientific discourse, but this does not imply failure on the part of science. Steinbeck, Ricketts, and Carson celebrate both the unknown dimension and the attempt to know it. We need both. More precisely, we need the point where they intersect. It follows, then, that the pragmatist ecology of *The Sea around Us* becomes an integration of text, observer, natural phenomena, and reader—a node of fact and lore, of doubt and inquiry that weakens the subject-object barrier and envisions all entities in ecosystemic relationships.

Her prose often celebrates these relationships, and Carson remains highly critical of the human destruction of ecosystems that goes hand in glove with our unexamined faith in science and reason. "Throughout *The Sea Around Us,*" writes Vera Norwood, "Carson points out humankind's inability to live in terms of the grand natural cycles science has enabled us intellectually, at least, to know" ("Heroines" 336). A good example is the chapter "The Birth of an Island," in which Carson describes the process of island building and comments on the human-induced tragedy of some Pacific islands. When ships first visited these isolated islands, Europeans introduced goats and cattle and exotic plants that decimated unique island ecologies. Even rats swimming ashore from a sinking ship could by themselves destroy an ecosystem, as had happened as late as 1943.[17] Writing about an Australian island where shipwrecked rats made land, Carson quotes an islander in an eerie echo of *Silent Spring:* "This paradise of birds has become a wilderness, and the quietness of death reigns where all was melody" (*Sea around Us* 94). Carson goes on to say: "The tragedy of the oceanic islands lies in the uniqueness, the irreplaceability of the species they have developed by the slow processes of the ages. In a reasonable world men would have treated these islands as precious possessions, as natural museums filled with beautiful and curious works of creation, valuable beyond price because nowhere in the world are they duplicated" (96). She inverts the dominant notion of *reason*. It is, of course, *reason* that leads people to *improve* islands by introducing agricultural animals and plants. Like Dewey, who complicates our common notions of use, Carson would revise our conception of reason so that it recognizes the value of ecological relationships and the sanctity of life-forms different from our own. We should treasure these things as we treasure great art, and recognize their aesthetic value, as she often does: "A hard, brilliant, coruscating phosphorescence often illuminates the summer sea," and "dolphins . . . fill the water with racing flames and . . . clothe themselves in a ghostly radiance" (*Sea around Us* 32).

Although Lawrence Buell finds that Carson equivocates the caustic bite of her chapter on islands in the rest of *The Sea around Us,* I see her view as more pessimistic (*Endangered* 200). In another icy echo of *Silent Spring,* Carson still retains hope for the inviolability of the sea: "There is the promise of a new spring in the very iciness of the winter sea" (*Sea around Us* 36). She remains on some levels caught up in the postwar optimism of the United States, praising advances in technology and oceanography spurred on by the war effort: "We seem on the edge of exciting new discoveries. Now oceanographers and geologists have better instruments than ever before to probe the depths of the sea, to sample its rocks and deeply layered sediments, and to read with greater clarity the dim pages of past history" (106–7). Carson was never opposed to technology per se but rather to unexamined—*reasonable*—faith in it: "Most of man's habitual tampering with nature's balance . . . has been done in ignorance of the fatal chain of events that would follow" (95). Even at this stage, she was deeply concerned about the consequences for the rest of creation as faith in science remained inviolable, as technology became powerful beyond all expectations, as rational knowledge and aesthetic perception became even more compartmentalized.

Its blending of science and art in a style accessible to a wide readership constitutes much of the appeal of *The Sea around Us.* This book cemented Carson's fame and enabled her to leave her job with the Fish and Wildlife Service and devote her time entirely to writing. It won the John Burroughs Award and the National Book Award, and the documentary film version won an Oscar. *The Sea around Us* proved a key text in the development of Carson's ecological aesthetic, and its fame gave her ample opportunity to speak about her creative process. On the occasion of accepting the National Book Award on 27 January 1952, Carson had much to say both about her aesthetics and the ethical valence of her writing practice. Concerning the poetic nature of her book, she claimed: "The winds, the sea, and the moving tides are what they are. If there is wonder and beauty and majesty in them, science will discover these qualities. If they are not there, science cannot create them. If there is poetry in my book about the sea, it is not because I deliberately put it there, but because no one could write truthfully about the sea and leave out the poetry" (qtd. in Lear 219). Like Thoreau, who memorably wrote, "Let us not underestimate the value of a fact, it will one day flower in a truth" ("Natural History" 130), Carson felt a certain poetry in facts. An accurate description of physical phenomena need not be mundane. Her poetry, she insists, is manifest in the factual sea itself and in her book. This is an approach different from imposing a narrative structure upon a natural entity, and it is not simply an extension of a nineteenth-century narrative of organic form. Perception brushes up against a world older than intelligence, and the intersection of the creative intelligence and the barely perceived world leads to aesthetic experience. The sea certainly predates human intelligence, yet we still carry the chemical structure of seawater in our very blood: "Fish, amphibian, and reptile, warm-blooded bird and mammal—each of us carries in our veins a salty stream in which the elements sodium, potassium, and calcium are combined in almost the same proportions as in sea water" (*Sea around Us* 13). Dewey seems to see a like connection: "Having the same vital needs, man derives the means by which he breathes, moves, looks and listens, the very brain with which he coordinates his senses and his movements, from his animal forbears. The organs with which he maintains himself in being are not of himself alone, but by the grace of struggles and achievements of a long line of animal ancestry" (*AE* 19). The structure and rhythm of the sea—of nonhuman nature—must, then, be in our veins, in our poetry, in our very being. This shared rhythm seems to me a prime example of the transcendent as the intersection of the tangible and intangible. As Linda Lear points out, Carson was, after all, a highly skilled writer and a tireless reviser, so it would be foolish to think that she was unaware of her own poetic skill (219). But Carson makes no claim to impose a Coleridgean imaginative order upon the sea.[18] On the contrary, her writing makes present in the world the ecotone where her literary skill and the poetry residing in the sea interpenetrate. As Lear so succinctly puts it, "Accuracy and beauty were never antithetical qualities in her writing" (219). Poised at an ecotone, Carson's words un-perch experience.

Carson's reference to "truth" in this speech is also telling. Truth inheres in relationships. Carson suggests that the words on the page bear truth, and again, as in the passages quoted earlier from *Under the Sea Wind,* it is not literal or photographic description that carries truth. Language that works to create a link between the human and the nonhuman world, whether by evoking an emotional connection to animals, a biological link to the primal ooze, or an aesthetic connection to the wonder of the sea, works to make present in the world the truth that many things are related and integrated through interlocking ecological webs. This truth also has an aesthetic realm. If we look again to Dewey, we read that "The thoroughgoing integration of what philosophy discriminates as 'subject' and 'object' (in more direct language, organism and environment) is the characteristic of every work of art. The completeness of the integration is the measure of its esthetic status" (*AE* 281).

This juncture between Dewey and Carson is intriguing in two ways. First, both insist that the aesthetic nature of a work of art is grounded in the quality of the integration of human observer/artist and the environing world. In their insistence Dewey and Carson anticipate more recent thinking about the central importance of relationality for ecologically inflected writing and philosophy. In writing about Terry Tempest Williams's *Refuge,* Mark Allister notes: "Her human bonds are certainly strong, but Williams looks outward, to the birds and the nonhuman world, to understand these connections, to give her peace and teach her about refuge" (66).[19] Anna L. Peterson argues that "Environmental ethics points to the further necessity of including nature or the land as an element of the moralscape, perhaps even as its underlying ground" (232). In the opening lines of *The Edge of the Sea,* Carson seems to anticipate this key focus on relations: "In the recurrent rhythms of tides and surf and in the varied life of the tide lines there is the obvious attraction of movement and change and beauty. There is also, I am convinced, a deeper fascination born of inner meaning and significance" (xiii). Inner value is sensed primarily in relation to the physical world, in the experienced integration of human and nonhuman.[20] The perception of relations is central to Carson's and Dewey's thought. Art and thought depend on the interrelation between the human and nonhuman worlds.

Second, but certainly not less important, both writers strive to make their work accessible to a larger reading public. Dewey's strategy is not at all subtle. He translates philosophical language for the lay reader parenthetically. Like Steinbeck and Ricketts, Dewey points with some sarcasm to the efforts of the philosophical community to keep its language abstruse and inaccessible.[21] But even more to the point, Dewey talks about interest and purpose: "All interest is an identification of a self with some material aspect of the objective world, of the nature that includes man. Purpose is this identification in action" (*AE* 281). Again, we clearly see Dewey anticipating more recent developments in relational thinking.[22] In order for pragmatist ecology to set experience in motion, there must be action taken on an individual's relation to the physical and cultural world, and the individual's full access to the art of knowing is in all our interest and essentially important to the continuity of experience.

Carson's work has this purpose integrated into its aesthetics. Science is one way of linking the self to the material world, as is creative activity. Carson employs both—through her art, she makes science available to a lay readership. In her acceptance speech, Carson reclaimed science for everyday experience:

> Many people have commented with surprise on the fact that a work of science should have a large popular sale. But this notion that "science" is something that belongs in a separate compartment of its own, apart from everyday life, is one that I should like to challenge. We live in a scientific age; yet we assume that knowledge of science is the prerogative of only a small number of human beings, isolated and priestlike in their laboratories. This is not true. It cannot be true. The materials of science are the materials of life itself. Science is part of the reality of living; it is the what, the how, and the why of everything in our experience. It is impossible to understand man without understanding his environment and the forces that have molded him physically and mentally. (qtd. in Lear 218–19)

Part of the enduring importance of Dewey, Carson, Steinbeck, and Ricketts is their attempt to make science and philosophy more accessible to a wider audience, while simply popularizing neither.

Science is "part of the reality of living," that is, part of experience; all four writers insist upon returning science and aesthetics to the realm of everyday experience. This insistence has aesthetic and ethical consequences, as Dewey points out in a passage in which he sounds uncannily similar to Carson in his call to retrieve science for common experience and to integrate science with aesthetics:

> Today the rhythms which physical science celebrates are obvious only to thought, not to perception in immediate experience. They

are presented in symbols which signify nothing in sense-perception. They make natural rhythms manifest only to those who have undergone long and severe discipline. Yet a common interest in rhythm is still the tie which holds science and art in kinship. Because of this kinship, it is possible that there may come a day in which subject-matter that now exists only for laborious reflection, that appeals only to those who are trained to interpret that which to sense are only hieroglyphics, will become the substance of poetry, and thereby be the matter of enjoyed perception. (*AE* 154)

Dewey's "rhythms," like Carson's, originate in natural rhythms such as "the ebb and flow of tides, the cycle of lunar changes, the pulses in the flow of blood, the anabolism and katabolism of all living processes" (*AE* 154). As Muir understood the rhythm of drops of rain, both Dewey and Carson perceive via art and science a linkage of self to the world and its rhythms. Science is linked to poetry, which in turn makes the rhythms of the world—of both science and poetry—available to "perception in immediate experience." Perception in immediate experience then becomes a key requisite to the art of knowing, understood as not only ecological but democratic. Dewey writes in "Creative Democracy—The Task before Us" about the need for access to information so necessary for a democratic culture:

Democracy is a way of personal life controlled not merely by faith in human nature in general but by faith in the capacity of human beings for intelligent judgment and action if proper conditions are furnished. I have been accused more than once and from opposed quarters of an undue, utopian faith in the possibilities of intelligence and in education as a correlate of intelligence. At all events, I did not invent this faith. I acquired it from my surroundings as far as those surroundings were animated by the democratic spirit. For what is the faith of democracy in the role of consultation, of conference, of persuasion, of discussion, in formation of public opinion, which in the long run is self-corrective, except faith in the capacity of the intelligence of the common man to respond with common-sense to the free play of facts and ideas which are secured by effective guarantees of free inquiry, free assembly and free communication? I am willing to leave to upholders of totalitarian states of the right and the left the view that faith in the capacities of intelligence is utopian. For the faith is so deeply embedded in the methods which are intrinsic to democracy that when a professed democrat denies the faith he convicts himself of treachery to his profession. ("Task" 227)

This passage has a special, sad ring to it in the face of contemporary U.S. politics, as it becomes starkly and painfully clear how deeply we need a renewed faith in the art of knowing in order to bring about a country and a world embedded in democratic and ecological values. A nexus forms here with Carson and Dewey. The key concept that binds them so closely together is their focus on everyday experience, and the aesthetic, ethical, political, and ecological ramifications of that focus. We perceive things within our experience, and too often, especially in Carson's (and our) context, that experience was (is) becoming increasingly unavailable to our everyday, lived experience through scientific specialization, which in turn inhibits the free play of facts and ideas.

Further, the free play of perception embedded in experience, as Dewey insists, is to be "enjoyed," so what is removed from the experiential realm is both the knowledge and joy that result from the everyday perception of the world. And the rewards of a democratic life. As seen in Dewey's parenthetical translation of philosophical terms, and as seen in all of Carson's work, this removal often happens when language becomes too specialized or falls into jargon, or whenever access to the art of knowing is curtailed. Dewey writes: "I am inclined to believe that the heart and final guarantee of democracy is in free gatherings of neighbors on the street corner to discuss back and forth what is read in uncensored news of the day, and in gatherings of friends in the living rooms of houses and apartments to converse freely with one another" ("Task" 227). Both common knowledge of science and uncensored news have been largely removed from our culture, compromising free inquiry. Both writers would redress this removal of the art of knowing from the public sphere. It is an ethical act that for both can be to some degree enacted through aesthetic means; again, art belongs in the common world, and "Artists have always been the real purveyors of news, for it is not the outward happening in itself which is new, but the kindling by it of emotion, perception, and appreciation"—of experience accessible to the public (*PP* 350). Norwood writes that "Carson's voice in all her books—her persona, if you will—is unequivocally allied with the nonspecialist" ("Knowing" 758), with the public. Both Dewey and Carson strive to create the "proper conditions" for a democratic culture rooted in ecological values. Carson desired to return both the discourse of science and the poetry of the natural world to the public. This desire permeates all of her work, and in her next book the integration of language and world—a pragmatist ecology—reaches new heights.

In Cheryll Glotfelty's estimation, after *The Sea around Us,* Carson "had become an authority whose opinions were quoted, rather than an unknown writer who quoted authorities" ("Carson" 158). Carson's next book, *The Edge of the Sea* (1955), is the key text in demonstrating the development of her ecological aesthetics and ethics. *The Edge of the Sea* is in some ways a companion piece to *The Sea around Us.* Where the earlier book attempts to grasp the entire sea, *The Edge of the Sea* engages the U.S. Atlantic shore and the three broad littoral habitats it provides: a rocky shoreline from Cape Cod north, a sandy one from Cape Cod south, and the coral reefs of the Florida Keys. She writes "about each geological area as a living ecological community rather than about individual organisms" (Lear 243). Each place is a life zone in itself, and each life zone contains myriad smaller divisions, each a world to itself, yet interrelated with all the others.

Houghton Mifflin first proposed *The Edge of the Sea* to Carson as a field guide, but she took the idea of a field guide and radicalized it (see Glotfelty, "Carson" 157; Lear 242–43). A field guide usually has sketches or photographs of birds or fish or trees followed by a simple description and tips toward field identification. Carson dismisses this procedure as inadequate. Things can be identified only in the context of their relationships to other animals and habitats. Her outlook is relational and ecosystemic: "Nowhere on the shore is the relation of a creature to its surroundings a matter of single cause and effect; each living thing is bound to its world by many threads, weaving the intricate design of the fabric of life" (*Edge* 14). Along with the recurrence of the tapestry allusion, this passage echoes this chapter's epigraphs from Leopold and Dewey, and in this book more than any other, Carson's aesthetics depends upon ecological principles of interrelation.[23]

Further, Carson's ecosystemic view of things has important ethical ramifications. In "Beyond Intrinsic Value: Pragmatism in Environmental Ethics," Anthony Weston writes: "What can exist and attract in isolation from everything else may be, for just that reason, *bad:* like the dream world of the drug user, it seduces us away from the complexity of our lives, substitutes solipsism for sociality, divides certain parts of our lives from the rest. We should prefer a conception of values which ties them to their contexts and insists not on their separability but on their relatedness and interdependence" (330). *The Edge of the Sea* has this ethical valence of ecology built into it. Carson asks a series of questions: "What is the message signaled by the hordes of diatoms, flashing their microscopic lights in the night sea? What truth is expressed by the legions of the barnacles, whitening the rocks with their habitations, each small creature within finding the necessities of its existence in the sweep of the surf? And what is the meaning of so tiny a being as the transparent wisp of protoplasm that is a sea lace, existing for some reason inscrutable to us—a reason that demands its presence by the trillion amid the rocks and weeds of the shore?" (*Edge* 250). While, of course, these creatures in their own world have no "message," no "truth," and no "meaning" as we conceive of these terms, they exist in a *relationship* that possibly does have an ethical message, truth, or meaning once that relationship is perceived by the creative intelligence and directed action follows—the writing of a book, for instance. Aesthetic experience is put in motion. Carson does not answer her rhetorical questions, because there are no answers, only a continuity of questions—the questions are open-ended and intended to contribute to ongoing experience, and in this way contingency makes an ongoing contribution to experience. The meaning inherent in these creatures is the unreachable meaning of life on the fringe of perception, and each creature contributes to the integrity of life in an interrelational, ecological community that includes the human being and its art. Meaning, again, inheres in relationships. Carson's identification with this community leads to an aesthetic purpose committed to the continuing interrelationship of all life.

William Chaloupka makes a similar claim for Dewey's aesthetics as a model for a purposeful environmental ethics. He contends that "when Dewey articulated the standards for worthwhile experience, he was in fact delineating the value of continuity, now familiar from the discourse of ecology, in a pragmatic, social context" (250–51).[24] For Dewey, this continuity is dependent upon the interaction of an organism with its environment. Continuity involves a doing and an undergoing; there is, in other words, a push and shove between the organism and the environment productive of experience. Dewey is also careful to note that this doing and undergoing is not limited to the physical environment. A person lifting a stone and one generating ideas are likewise engaged with an environment. The ongoing engagement generates experience. However, experience is subject to denigration when there is an excess of doing or undergoing. For example, if I watch television all day or never take off the earphones of my *iPod* except to sleep, I undergo too much—experience has no time to develop. If I do nothing but work all day every day, I do too much—in other words, I don't pay attention. Without conscious attention, experience cannot happen: experience depends upon the perception of the relations between doing and undergoing. This perception of relations is a form of ecological connection to the world, and Chaloupka adds, "Not only is everything connected, as the ecologist might put it, but that connection is somehow centrally important, not a curiosity which is occasionally of real interest" (251).

Many passages in *The Edge of the Sea* are perceptions of these centrally important connections and relations. Carson's prose gears these perceptions to the physical environment and engages the reader with the ecology of facts and ideas contained in the book. Paul Brooks insists that "As a writer she used words to reveal the poetry—which is to say the essential truth and meaning—at the core of any scientific fact. She sought the knowledge that is essential to appreciate the extent of the unknown" (*House of Life* 7). *The Edge of the Sea* is an ecotonal text in which the mental, physical, known, and unknown worlds interrelate. At the beginning of the book, looking out over a cove on the western coast of Florida, Carson senses that

> The sequence and meaning of the drift of time were quietly summarized in the existence of hundreds of small snails—the mangrove periwinkles—browsing on the branches and roots of the trees. Once their ancestors had been sea dwellers, bound to the salt waters by every tie of their life processes. Little by little over the thousands and millions of years the ties had been broken, the snails had adjusted themselves to life out of water, and now today they were living many feet above the tide to which they only occasionally returned. And perhaps, who could say so many ages hence, there would be in their descendants not even this gesture of remembrance for the sea. (6–7)

For most people, perhaps, the snails on the mangrove roots would qualify in Chaloupka's terms as a "curiosity" or of occasional interest, but Carson is able to translate their silent experience into a general experience of time and evolution, making it both centrally important and available to her readers. The snails are more than a "curiosity." To adulterate a bit Dewey's comment quoted earlier, "through art, the meaning of" the snails "that are otherwise dumb, inchoate, restricted, are clarified and concentrated, and not by thought working laboriously upon them, nor by escape into a world of mere sense, but by creation of a new experience" (*AE* 138). We saw earlier that Carson in *The Sea around Us* insisted that the physical human being retains some elements of the sea in his or her body, and that through this subtle link we can see the relationship of our own evolution to the evolution of the snails. We share a similar pattern of adjustment. Dewey suggests that under the influence of Darwinian logic we become less interested in who or what made the world than in what kind of world it is, and less concerned with absolutes than with particular intelligences and the relations among them ("Darwinism" 11). In these relations, all is in fecund transition; we—snails and humans—are linked, and that link is "clarified" and "concentrated" and centrally important.

And Carson attends to the perception of further relations. Her experience does not stop with her thoughts on the snails in isolation—it has continuity.[25] Spurred on by horn shells scattered along the beach, once the primary food of flamingos, Carson desires: "I might see what Audubon saw, a century and more ago. For such little horn shells were the food of the flamingo, once so numerous on this coast, and when I half closed my eyes I could almost imagine a flock of these magnificent flame birds feeding in that cove, filling it with their color. It was a mere yesterday in the life of the earth that they were there; in nature, time and space are relative matters, perhaps most truly perceived subjectively in occasional flashes of insight, sparked by such a magical hour and place" (*Edge* 7). In a flash of insight spurred by a small shell lying in the sand, Carson is able to imagine the beach as a swash of living color. Her vision is rendered even more compelling by the juxtaposition of the evolutionary scale of millions of years enjoyed by the periwinkles, creatures that human beings have left alone, to the tragic fate of the flamingos, an animal too beautiful for human beings not to have slaughtered in huge numbers. One feels a deep sense of loss, and the birds are raised to almost Homeric status through the epithet "flame birds." Her use of phrases such as "might see," "half closed my eyes," and "could almost imagine" lend a special poignancy to the sense of loss, as if the birds and the beauty they brought with them were just out of reach. In the last sentence of the quoted passage, Carson also makes an interesting connection between the particular and the general by juxtaposing "time and space" to "hour and place."

Dewey staked his faith in democracy on the local community, and for Carson here, to undergo a profound experience in a particular hour and place—to experience the local—is to experience the world in its largest categories: time and space.

Carson perceives relationships here in an ecological web that connects time and space, abstract and concrete, ocean and shore, present and vanished creatures, and past and present writers and painters of the natural world. In fact, the distinctions between these pairs begin to vanish, and dualistic thinking begins to drop away. Dewey writes: "Experience is limited by all the causes which interfere with perception of the relations between undergoing and doing" (*AE* 51). Primarily, figuring oneself—whether artist, critic, scientist, or philosopher—as a subject gazing onto an objectified world elides the possibility of experience. It denies the subject's intimate and intricate relation to the object-world, or aesthetic saturation in Dewey's terms: "Saturation means an immersion so complete that the qualities of the object and the emotions it arouses have no separate existence" (*AE* 280). In a recent study of material culture in American literature, Bill Brown seems to echo Dewey when he writes of Sarah Orne Jewett's *The Country of the Pointed Firs* that "The culture of nature, as she describes it, is one where nature comes to saturate bodily life" (90). The duality of things and human life falls away. As in her famous predecessor's novel, in Carson's prose we are able to perceive the fact that culture and the natural world in all their ramifications exist in an ecotonal relationship; there is a saturation of undergoing and doing on a cultural/natural level. If we refer back to Weston's construction, we see here in Carson an aesthetics with a purposeful ethics of "relatedness and interdependence."

The Edge of the Sea, more than the two books preceding it, allows Carson's personal experience to saturate its discourse. The book is framed by Carson's personal observations and thoughts about the sea. This adds another layer to the texture of the book, but more importantly, it takes the human observer out of the background and foregrounds the human author in a position relative to her environment as both subject and object. Carson attempts to see objectively, but her own response to the world is a subject of the book. This integrates the human response into the ecosystem of the shoreline. For example, the first sentence of the final chapter takes us both to the sea and to the site of writing: "Now I hear the sea sounds about me; the night high tide is rising, swirling with a confused rush of waters against the rocks below my study window" (249). The writing subject and the subject of her writing interpenetrate; in Dewey's terms, there is an integration of organism and environment, a saturation. The sounds of the sea, the tides and water and rock find their way into Carson's study and onto the page—they are literally there with her. She is implicated in a web of relationships that interlaces the physical and the textual, and such implication marks experience that "signifies active and alert commerce with the world; at its height it signifies complete interpenetration of self and the world of objects and events" (*AE* 25). The seashores are perhaps the most productive ecotones on earth, and Carson's book, itself ecotonal, installs her experience in nature, embedding this rich, relational zone on the margin of the sea in public space accessible to all. In this way, the book draws its author, the physical world, and its readers into the height of ongoing ecological experience.

The Edge of the Sea opens with an introductory section called "The Marginal World," in which Carson opens her experience to her readers:

> Once, exploring the night beach, I surprised a small ghost crab in the searching beam of my torch. He was lying in a pit he had dug just above the surf, as though watching the sea and waiting. The blackness of the night possessed water, air, and beach. It was the darkness of an older world, before Man. There was no sound but the all-enveloping, primeval sounds of wind blowing over water and sand, and of waves crashing on the beach. There was no other visible life—just one small crab near the sea. I have seen hundreds of ghost crabs in other settings, but suddenly I was filled with the odd sensation that for the first time I knew the creature in its own world—that I understood, as never before, the essence of its being. In that moment time was suspended; the world to which I belonged did not exist and I might have been an onlooker from outer space. (5)

Alone with the crab on the night beach on the Georgia coast, Carson has an epiphany of sorts. She shares an eloquent awareness that although humans are related to all things, we remain emphatically different. This awareness surfaces periodically in her writing, but this time Carson draws readers into her own, very real discomfort—into "the darkness of an older world"—and then she is suddenly filled with the wonder of new discovery in a scene she had encountered hundreds of times before. She does not claim to know or understand, but to have the "odd sensation" that she does both, and this only in a fleeting moment. This is clearly a brush between the tangible, the human, and the intangible, the world of the crab. In this moment, even Carson's sense of self is loosened and it seems as if her consciousness and the crab world saturate one another. This does not remove the crab to a transhistorical or transcendental realm, but rather momentarily approaches transcendence in Dewey's sense as Carson finds herself in the interstices of her world and the crab's world. This seems a redemptive moment to me, partially in the nature of things, partially the result of philosophical agency. Carson undergoes a shiver of realization that she somehow knows the world of the crab, and this in turn throws her world into question. She finds herself in a realm of intersubjectivity and experiences herself not as an isolated ego but as a participant among other subjectivities in an ecosystemic interrelationship. She insists that any ecological aesthetic or ethical structure strive to respect the others with whom we participate in this world, and this respect entails relinquishing our belief that we stand separate from the physical environment. We are among myriad subjectivities. As we diminish these others we do two things: first, we deny other beings the right to evolve in their own way; and second, we degrade the texture of the world for ourselves and all living things. A purely human world would be profoundly void of the ecotonal spaces where both life and art are generated.

As mentioned earlier, in Carson's work the surface of the sea itself makes visible the idea of an ecotone, and in *The Edge of the Sea* the idea becomes far more complex. Carson writes: "For it is now clear that in the sea nothing lives to itself. The very water is altered, in its chemical nature and in its capacity for influencing life processes, by the fact that certain forms have lived within it and have passed on to it new substances capable of inducing far-reaching effects. So the present is linked with past and future, and each living thing with all that surrounds it" (37). The seawater drives the current of life itself, and all forms of living things contribute to the power of the water to support life.[26] The sea provides an interface not only of creatures and habitats but of time itself. It contains the past, present, and future; in the cycle of things, the distinctions between these humanly defined temporal categories become blurred. Time no longer seems linear—if anything it is cyclical, linked with the contingencies and processes of living, dying, and decomposing of various lifeforms. To take this cycle one step further, the creatures of the coral reefs blur our distinctions between biology and geology, between living and non-living things:

> All are important in the economy of the marine world—as links in the living chains by which materials are taken from the sea, passed from one to another, returned to the sea, borrowed again. Some are important also in the geologic processes of earth building and earth destruction—the processes by which rock is worn away and ground to sand, by which the sediments that carpet the sea floor are accumulated, shifted, sorted, and distributed. And at death their hard skeletons contribute calcium for the needs of other animals or for the building of the reefs. (*Edge* 221–22)

The building of the earth happens within the sea, and the importance of these little creatures begins to rival the efforts of human beings. For all our monumental construction projects, the tiny creatures of the sea are essential to building the world itself. They contain the potential for its future. Carson's assertions here are admittedly subtle, but we begin to sense that we destroy these other worlds only at dire risk to ourselves. Along with the pleasure derived from and the beauty perceived in Carson's writing comes an ethical imperative that we handle ourselves with renewed respect and restraint toward the natural world that we barely understand.

There are, of course, many reasons why we should behave respectfully toward the natural world. Many of them are practical—we may run out of wood and water—but Carson makes other reasons clear too. Part of her project seems to be the reconstitution of our sense of pleasure in the world in which we occur. Simply, we are fortunate to be living in an often marvelous world, participants in a complex, ongoing process of life on earth about which we have no final knowledge. Carson takes this concept of interrelationship out of the laboratory, where, as Buell comments, "These arenas of biological interdependence can . . . be talked about in wholly clinical ways devoid of political or affective content" (*Environmental Imagination* 302).[27] She casts the scientific fact of interrelationship in "affective," aesthetically pleasing prose. An aesthetic appreciation of anything usually entails an enjoyment of that thing, and in renewing readers' capacity for joy and wonder in their interdependence with the biosphere, Carson takes an aesthetic and ethical position toward both human and other-than-human environments. Humans as a species cannot, any more than Dewey's democratic individual, exist alone. We will thrive only in ecological, democratic communities.

Carson's stance lines up with Dewey's aesthetics and the ethical stance it too implies. At the end of *Art as Experience*, Dewey writes: "Art is a mode of prediction not found in charts and statistics, and it insinuates possibilities of human relations not to be found in rule and precept, admonition and administration" (352). For Dewey, art enacts and at the same time is everyday experience of the interaction of an organism and its environment. Art predicts "possibilities," shows us the actual in the light of the possible. It does not work toward specific ends but rather toward a continuity of experience. Revising the dominant, bureaucratic scientific discourse of her time, Carson charts scientific facts and concepts in literary language. For example, *The Edge of the Sea* includes an appendix "for the convenience of those who like to pigeonhole their findings neatly in the classification schemes the human mind has devised" (xiv). Carson's language is clearly dismissive of neat classification, and she insists that "To understand the shore, it is not enough to catalogue its life. Understanding comes only when, standing on a beach, we can sense the long rhythms of earth and sea that sculpted its landforms" (xiii). If we recall, Dewey insisted that "a common interest in rhythm is still the tie which holds science and art in kinship," and he hopes to see science and art bound together in experience (*AE* 154). Aesthetic experience predicts deeper human understanding and depends upon the perception of the relations between undergoing and doing, between the earth and the sea, between the human and the land. Looking into a tide pool "paved" with mussels, Carson sees that "The water in which they lived was so clear as to be invisible to my eyes; I could detect the interface between air and water only by the sense of coldness on my fingertips. The crystal water was filled with sunshine—an infusion and distillation of light that reached down and surrounded each of these small but resplendent shellfish with its glowing radiance" (*Edge* 115). The ecotone here between air and water is invisible yet cool to the touch, and Carson does not insist that we *see* this permeable boundary, but that we *look* at it not only with our eyes but also with all our senses and our imaginations, that we *feel* it. It is an ethical commitment of the human organism to looking at the world beyond itself. It is a commitment to experience.

This commitment does not release us from the bonds of language and culture. Carson employs metaphors deeply ingrained in the Western cultural imaginary to evoke a sense of wonder at the world into which she looks. The water is like a crystal chalice, full to the brim, in which light is infused and distilled, and the light "reaches down" to the mussels in an image packed with religious overtones. The light is like the hand of a god, and the mussels are "resplendent" in its "radiance," haloed if you will. Further, the mussels provide a site of connection, and Carson, with her metaphor, prepares the reader to perceive what she seems to insist is a level of sacredness in the ecological concept of interconnection: "The mussels provided a place of attachment for the only other visible life of the pool. Fine as the finest threads, the basal stems of colonies of hydroids traced their almost invisible lines across the mussel shells" (*Edge* 115). Carson extends the metaphor of the "crystal water" to the description of the hydroids, whose "branches are enclosed within transparent sheaths, like a tree in winter wearing a sheath of ice" (115), linking, as she did in "Undersea," the ecosystem of the tide pool to the ecosystem of the land. She continues: "From the basal stems erect branches arose, each branch the bearer of a double row of crystal cups" (115–16), and the metaphor is extended further with the connection to the cup-bearer, one who holds a profoundly important place in religious and heroic lore.

Of course, the hydroids are predators, and because Carson cannot see any other life-forms in the water it does not mean they are not there. In fact, they must be there: "Somewhere in the crystal clarity of the pool my eye—or so it seemed—could detect a fine mist of infinitely small particles, like dust-motes in a ray of sunshine. Then as I looked more closely the motes had disappeared and there seemed to be once more only that perfect clarity, and the sense that there had been an optical illusion. Yet I knew it was only the human imperfection of my vision that prevented me from seeing those microscopic hordes that were the prey of the groping, searching tentacles I could barely see" (116–17). Carson, one could say, is "groping" and "searching" too. The "crystal" metaphor is carried through, yet in spite of the limpid environment, Carson can "barely see." With this construction, "barely see," she sets the human being in relationship to the small ecosystem she has perceived in the tide pool. She realizes that she is not (or barely) a part of this little world, yet by casting it in affective prose she draws herself and her readers into the life of the tide pools. There is something literally at the fringe of her sight that initiates the perception of relations and sets experience in flight. However, we can read here a note of caution. This world is something we can barely sense; for Carson, it is sacred in that it is an example of the ecological intricacy of life itself, yet the human apparatus can barely perceive it, cannot, certainly, understand it in its full complexity. In the face of such beauty and complexity, our approach should be one of restraint. We damage these places and creatures to the detriment of experience itself.

Carson distills her aesthetic and ethical practice when she recalls her first sighting of a West Indian basket star: "For many minutes I stood beside it, lost to all but its extraordinary and somehow fragile beauty. I had no wish to 'collect' it; to disturb such a being would have seemed a desecration" (*Edge* 225). Her experience with the starfish provides a marker for an ecological stance toward the entire biosphere. The star is not something for humans to "collect," though we have the right to enjoy it as long as we leave it as undisturbed as possible. Its beauty is wonderful, and we know theoretically that it and its habitat are extremely fragile, yet we cannot seem to put our theoretical knowledge to practical use. Alone, theory is little help. Again, the allusion to religion with the term "desecration" suggests that the desecration of the natural world is, for Carson, akin to the "desecration" of a temple. At its most radical sense—at its root— Carson's call is for respect and restraint concerning the natural world.

Carson's ethical imperative, in large part grounded in aesthetic appreciation, has an unpalatable ring in our dominant culture.[28] Just as Thomas Lyon did, Paul Brooks calls Carson's ethical and aesthetic position a "heresy." He writes that in her stance toward the scientific establishment she declared "the basic responsibility of an industrialized, technological society toward the natural world. This was her heresy. In eloquent and specific terms she set forth the philosophy of life that has given rise to today's environmental movement[s]" (*Speaking for Nature* 285).[29] This philosophy is grounded in the interrelatedness of all things. To forge a theoretical link here to pragmatist thought, Weston argues that "Pragmatism insists most centrally on the *interrelatedness* of our values. The notion of fixed ends is replaced by a picture of values dynamically interdepending with other values and with beliefs, choices, and exemplars: pragmatism offers, metaphorically at least, a kind of 'ecology' of values" (322). For Dewey, value is rooted in the recognition that "this human situation falls wholly within nature. It reflects the traits of nature; it gives indisputable evidence that in nature itself qualities and relations, individualities and uniformities, finalities and efficacies, contingencies and necessities are inextricably bound together" (*EN* 314). Value is a process embedded in nature. At the close of *The Edge of the Sea*, Carson writes: "Once this rocky coast beneath me was a plain of sand; then the sea rose and found a new shore line. And again in some shadowy future the surf will have ground these rocks to sand and will have returned the coast to its earlier state. And so in my mind's eye these coastal forms merge and blend in a shifting, kaleidoscopic pattern in which there is no finality, no ultimate and fixed reality—earth becoming fluid as the sea itself" (249–50). The state of the world is as fluid as Thoreau's sandbank. Rock, sand, earth, and water are in flux. There is no final reality; all is cyclical and contingent, just as is a pragmatist approach to valuing experience. The world Carson perceives is ecological; it has no narrative, and it is no fable. It is a matter of patterns in which all things are interrelated, connected, and the effort to enjoy experience is the effort to enact continuity in the sense that Chaloupka suggests we use the term. What enhances the integrity of the natural world and the integrity of human experience is what encourages continuity—not continuity aimed toward a specific end, but toward ongoing experience woven into an ecologically intact community.

By the time *The Edge of the Sea* was published, people were beginning seriously to worry about the possibility of nuclear warfare and the threat it posed to all living things. The looming threat to intact ecosystems had never been so apparent. There was at large another danger to living things: the widespread, government-sponsored use of pesticides such as DDT had begun to have noticeable effects on domestic animals and wildlife across the United States. The chemicals entered food chains, and by the time they reached songbirds, for example, they were concentrated in lethal doses in the tissues of the animals. Bird eggs were so frail that birds were becoming unable to reproduce. The evidence was overwhelming, and in 1958 Carson writes in a letter to her friend Dorothy Freeman:

> It was pleasant to believe, for example, that much of Nature was forever beyond the tampering reach of man—he might level the forests and dam the streams, but the clouds and the rain and the wind were God's. . . .
>
> It was comforting to suppose that the stream of life would flow on through time in whatever course that God had appointed for it—without interference by one of the drops of the stream—man. And to suppose that however the physical environment might mold Life, that Life could never assume the power to change drastically— or even destroy—the physical world. (qtd. in Lear 310)

It had become impossible to believe this. The formerly positive interpretation of interrelatedness and interconnection has, of course, a negative side, and a growing awareness of the fragility of the natural world provides the context in which Carson began to work on *Silent Spring*.

Interrelationship as a key term for ecological thinking underwent a semantic shift from one that generally indicated a sense of well-being to one that included a dire threat to life itself, and in Carson this shift provoked a radical sense of responsibility toward life.[30] Carson's books up to this point develop an ecological aesthetic in which all things are interrelated. It is not a far step from this aesthetic perception to ethical consideration for nonhuman things. Carson believed that the world in all its beauty was dependent on the integrity of all creatures, and when her faith that the physical world itself was threatened by a bureaucracy seemingly bent on maniacal self-destruction, she turned her pen most explicitly to advocacy for life itself. Her advocacy elicited a vehement response. "Over the course of the controversy," notes Linda Lear, "it became clear to her enemies as well as her allies that Carson had forced a public debate over the heretofore academic idea that living things and their environment were interrelated" (429). To incessantly tug this notion of interrelation out of the realm of theory and into the world of practice constitutes another heresy—this is the role of pragmatist ecology. We should keep it firmly in mind.

It is beyond the limits of this chapter to say much more about *Silent Spring* itself. That would take another study. Carson's ecological perception of the sea-world that she aligned with concepts such as truth and poetry allowed her to see the entire world as a sacred place and finally to recognize and denounce the power we have attained to destroy vast ecosystems, perhaps, to her mind, the world itself, because a world without birds or fish or insects was, for her (and for us), no world at all. She asked a question that clearly and lyrically links ecology and democracy: "Who has decided—who has the *right* to decide—for the countless legions of people who were not consulted that the supreme value is a world without insects, even though it be also a sterile world ungraced by the curving wing of a bird in flight?" (*Silent Spring* 127, Carson's emphasis).

Just as she began *The Edge of the Sea* with a personal anecdote of a small person in a large world, Carson begins *Silent Spring* with a fable about any small town in a big country. In this case, the rhetorical ploy of the fable is apt. She uses it to cement her largely suburban readership—to instill an awareness of and to awaken a fear for the dangers posed to their own communities by pesticide use.[31] She sets up an idyllic landscape, "There once was a town in the heart of America where all life seemed to live in harmony with its surroundings" (1), and introduces into it silence and poison:

> There was a strange stillness. The birds, for example—where had they gone? Many people spoke of them, puzzled and disturbed. The feeding stations in the backyards were deserted. The few birds seen anywhere were moribund; they trembled violently and could not fly. It was a spring without voices. On the mornings that had once throbbed with the dawn chorus of robins, catbirds, doves, jays, wrens, and scores of other bird voices there was now no sound; only silence lay over the fields and woods and marsh. . . .
>
> In the gutters under the eaves and between the shingles of the roofs, a white granular powder still showed a few patches; some weeks before it had fallen like snow upon the roofs and the lawns, the fields and streams. (2, 3)

Carson brings awareness and fear into the homes of average Americans, and she makes present the threat to perhaps the most cherished American symbols of that time, the home and the local community.[32] The rhetoric of *Silent Spring*, unlike that of the books preceding it, calls for immediate political action, surely because the threat to life was so imminent, and perhaps partly because Carson knew she was dying of breast cancer. Her rhetoric was profoundly effective, and she met with rabid opposition.

As I have said, this book went on to change the way the population of the United States perceived the role of government and industry in their democracy. Because of this, Carson was bitterly attacked on many fronts. One letter reads:

> Miss Rachel Carson's reference to the selfishness of insecticide manufacturers probably reflects her Communist sympathies, like a lot of our writers these days.
> We can live without birds and animals, but, as the current market slump shows, we cannot live without business.
> As for insects, isn't it just like a woman to be scared to death of a few little bugs! As long as we have the H-bomb everything will be O.K. PS. She's probably a peace-nut too. (qtd. in Lear 409)[33]

Aside from all else, the writer obviously never bothered to read the book. And one chemical producer felt so threatened that it produced a parody of the fable quoted above and distributed it to newspapers all over the

United States: "Quietly, then, the desolate year began. Not many people seemed aware of danger. After all, in the winter, hardly a housefly was about. What could a few bugs do, here and there? How could the good life depend upon anything so seemingly trivial as bug spray? Where were the bugs anyway? The bugs were everywhere. Unseen. Unheard. Unbelievably universal. Beneath the ground, beneath the waters, on and in limbs and twigs and stalks, under rocks, inside trees and animals and other insects—and, yes, inside man" (qtd. in Lear 431). I will end by saying that the parody came from Monsanto, a corporation that has abandoned chemicals in favor of genetically modified crops and seeds and their aggressive global marketing. We have not grown into Carson's ethics. And, again, it seems reasonable to ask now if it is not wise to grab Carson's legacy from out of the symbolic realm and begin to put it back into use.

Former Richardson Roadhouse, seventy miles south of Fairbanks on Richardson Highway. Courtesy of Paul Greimann Collection, Archives, Alaska and Polar Regions Collections, Rasmuson Library, University of Alaska, Fairbanks.

4
"The Coldest Scholar on Earth"

Silence and Work in John Haines's
The Stars, the Snow, the Fire

Arctic history became for me, then, a legacy of desire—the desire of individual men to achieve their goals. But it was also the legacy of a kind of desire that transcends heroics and which was privately known to many—the desire for a safe and honorable passage through the world.

—Barry Lopez

John Haines is a man of character in the most American sense—solitary; strong. He has spent much of his life in wild country, and he has written of the wilderness with profound wonder, perception, and thanksgiving. He has given us an indispensable sense of our country, our continent, and our earth in his poems. His work, his whole work, enriches us.

—1990 Western States Book Awards Jury Citation

While Rachel Carson was studying and writing about the sea, John Haines was crafting poems and, possibly, making a harness for his sled dogs from the hide of a moose he shot to keep himself fed through the winter. He could have been checking his trap lines. Perhaps he was gathering mushrooms or berries, cutting firewood, or cooking and scraping a porcupine for his dogs. He may just as well have been reading Virgil, Hermann Broch, or William Carlos Williams. In his collection of essays *The Stars, the Snow, the Fire* (1989), Haines looks back on his years homesteading near Richardson in the Alaska interior. He writes about an "honorable passage through the world," about "profound wonder, perception, and thanksgiving." Many people have written about the wilderness experience in North America, fewer have lived it, and fewer still, far fewer, have experienced it and possess the voice of a poet to express that experience. In this sense, Haines's book is a gift, a last song about a type of experience that is arguably no longer available to us. Haines first went to the interior of Alaska in 1947 and stayed for much of twenty-five years, his longest period of residence lasting from 1954 until 1969. His first book of poems, *Winter News*, published when he was forty-two, gained national attention in 1966.

In his prose, Haines, like Muir, documents the coming to awareness over time of a uniquely talented individual in wild country. Unlike Muir, Haines lived and worked in the country for an extended amount of time, and in this he stands out from the other writers considered in these pages. Haines, to be sure, as the others, focuses on the interrelationships between the human and the physical world, but in *The Stars, the Snow, the Fire* the physical environment, the writing, and the life expressed by the writing interrelate most often through the representation of basic physical work. This is not to say that Rachel Carson and John Steinbeck did not spend difficult hours working in the littoral zones of the Atlantic and Pacific coasts of North America, or that writing for a living is not work—it certainly is. But the work that connects Haines to the physical world in *The Stars, the Snow, the Fire* engages the world in a more primal way, and for Haines that work, at least prior to its representation in text, was a matter of survival. But I also feel safe in saying that this same work textualized is also a matter of aesthetic survival for Haines. In a valuable study of the georgic tradition in early American literature, Timothy Sweet argues that "Philosophical accounts outside the georgic tradition have often addressed the question of our relation to nature through the category of perception. Yet so long as our focus remains perceptual, what we actually do in the world and how we do it may remain secondary issues" (155–56).

I agree that our actions in the world merit the highest attention, but in the work of Haines and Dewey the terms of Sweet's argument seem to be reversed. For Haines and Dewey, it seems that most often, work enables perception, or, better, work can be a kind of perception. In Dewey's aesthetics, work and everyday experience help create the conditions for the perception of relations, hence for aesthetic value:

> In order to *understand* the esthetic in its ultimate and approved forms, one must begin with it in the raw; in the events and scenes that hold the attentive eye and ear of man, arousing his interest and affording him enjoyment as he looks and listens: the sights that hold the crowd—the fire-engine rushing by; the machines excavating enormous holes in the earth; the human fly climbing the steeple-side; the men perched high in air on girders, throwing and catching red-hot bolts. The sources of art in human experience will be learned by him who sees how the tense grace of the ball player infects the onlooking crowd; who notes the delight of the housewife in tending her plants, and the intent interest of her Goodman in tending the patch of green in front of the house; the zest of the spectator in poking the wood burning on the hearth and in watching the darting flames and crumbling coals. (AE 10–11, Dewey's emphasis)

The aesthetic is the scene of engagement, is a matter of participation. It is only understood by one who can perceive the relations between ballplayer and crowd, between husband and wife and their suburban plot of land. Workers high in the air are engaged in their own ballet of labor, but the aesthetic nature of their work remains unfulfilled until someone (even themselves) perceives the relations among the workers, their work, and the cultural and physical environment. It becomes now difficult to separate work from perception, and this move is typical of Dewey in his effort to escape dualistic thought and consider aesthetic experience as a continuity.

Dewey is especially interesting in this passage for what he includes in the category of aesthetic, and the all-encompassing nature of his aesthetics also resonates deeply with a democratic culture, which in its Deweyan construal is widely inclusive. The aesthetic is clearly a matter of active relationships. There is in this passage very little passive appreciation. Of course, the audience itself as an active participant completes the aesthetic equation. Dewey goes on: "The intelligent mechanic engaged in his job, interested in doing well and finding satisfaction in his handiwork, caring for his materials and tools with genuine affection, is artistically engaged," just as artists in their studios are engaged (*AE* 11). Although Dewey here draws strongly on urban and suburban imagery, the contention that work and quotidian life provide a source of aesthetic connection to both physical and cultural worlds is without doubt. It is clear in Haines's work as well.

This sense of connection—in Haines the sense of the physical world as a primary participant in human work—largely accounts for the fact that, as Kevin Bezner writes, "The true breakthrough in Haines's career came with the publication of a memoir, *The Stars, the Snow, the Fire*, which examines his years in Alaska" ("John Haines" 276). Marty Cohen writes of *The Stars, the Snow, the Fire* that "it's a book that will sit prominently next to the Northern writings of Jack London, Rockwell Kent, Farley Mowat, and Barry Lopez; considered strictly for its prose style, it's probably stronger than any of these" (145). But there is something else about Haines's writing that strikes Cohen as soon as the inevitable comparison to Robert Bly comes up: when one looks from Bly's to Haines's lines, "One feels corrected: Haines's language is literal, the experience he describes surreal or imagistic only in comparison with the denatured experience of most writers—even with most Northwestern writers" (146).[1] It is this renatured sense of connection to and participation with the natural world that we can best take from Haines and put to use. Already recognized as a fine poet, with the publication of this book of essays Haines emerges as a major voice in contemporary American nature writing.[2]

Although profoundly moved by the idea and experience of wilderness, Haines is, however, focused less on external nature than on what it means for a human being to live and work in close contact with the physical world. The aesthetic value in this book is lodged in a particular way of inhabiting a particular place. His connection to community and place is the source of his art. Thus, art becomes political in large part because, as Dewey teaches, the local is the source of democracy. Through his art, his place becomes a world, and Haines's way of living in it becomes a pattern for possible ways of living in the world generally. We learn in the course of the book that Richardson was at one time more populated than it was at the time of Haines's residency and that the landscape has been altered by miners, trappers, and pipeline crews. It is a wild place, certainly, but it is also deeply inscribed with human presence, and in this way the land becomes an important nexus of nature and culture. Full of ice, blood, bone, fur, death, hardship, and especially silence, Haines's world is by no means idyllic. His particular experience in it is brought back to bear in ongoing experience through its textualization. Experience is made available to the culture at large—in a sense, made universal—through reflection on a past that is brought into the present and the future through its articulation. And, to recall Derek Attridge's argument cited earlier, such an articulation "may alter cognitive frameworks across a wider domain, allowing further acts of creativity in other minds" (23).

Attridge visualizes a creative work as a "field of potential meaning," and his argument calls for a responsible stance toward the other. A response to the other—either as text or person—"will involve a suspension of my habits, a willingness to rethink old positions" (25). Such an ethical response allows for creativity and a respectful attitude toward the other, without which/whom the complexity of the self necessary for creativity would vanish. This idea of creativity is extended to all planes of thought and expression, of which the ethical is primary. As I argued earlier, one aspect of pragmatist ecology is its ethical consideration of the natural world as a participant in experience, and this creative ethics is basic to Haines. For instance, walking along the Richardson road in winter, he finds a frozen redpoll that triggers recollection and reflection: "In that tiny, quenched image of vitality, a bird like a leaf dropped by the wind in passing, I felt something of our common, friable substance—a shared vulnerability grasped once with insight and passion, and then too easily forgotten" (*Stars* 121). A hedge against forgetfulness, Haines's prose brings the shared fragility of human and nonhuman life to light.

And this reflection, we recall, far from the search for some foundation, becomes a recognition of interrelationships. As Dewey insists, "Everything that exists in as far as it is known and knowable is in interaction with other things" (*EN* 138). The physical world is one of those "other things": "In the degree, however, in which the mind is weaned from partisan and ego-centric interest, acknowledgement of nature as a scene of incessant beginnings and endings, presents itself as the source of philosophic enlightenment" (83). The movement from egocentric to ecocentric allows us to perceive the physical world as a process, a world of flights and perchings. The human being is continuous with the physical world; aesthetic experience depends upon this continuity. Recognizing the participatory relation between the human and the world is at the heart of pragmatist ecology, and the best ecological writing can aid us in this recognition. Again, Dewey's work underpins this train of thought: "the experiences that art intensifies and amplifies neither exist solely inside ourselves, nor do they consist of relations apart from matter. The moments when the creature is both most alive and most composed and concentrated are those of fullest intercourse with the environment, in which sensuous material and relations are most completely merged" (*AE* 109). This passage recalls the previous chapter's discussion about Dewey's idea of aesthetic saturation, in which natural objects so saturate human physical and emotional life that distinctions between the categories dissolve. Ego to eco, boundaries fall away, and the experience of the environment becomes one in which the entire realm of the senses is involved. Experience becomes aesthetic, ecological.

Another key strain in Dewey's argument asserts that art belongs to the common world: "The times when select and distinguished objects are closely connected with the products of usual vocations are the times when appreciation of the former is most rife and keen. When, because of their remoteness, the objects acknowledged by the cultivated to be works of fine art seem anemic to the mass of people, esthetic hunger is likely to seek the cheap and vulgar" (*AE* 12). This insistence that art belongs to the common world, that art should not be the province of an elite, emphasizes a primary goal of Dewey's philosophy—the recovery of democracy—in which art plays a central role toward "recovering the continuity of esthetic experience with normal processes of living" (16). Dewey calls to make art an intricate part of everyone's life—to make of life an art. Meditating on a frozen redpoll and physical work are examples of these processes, of creating aesthetic experience from common life and its interrelationship with the physical world.

Haines's prose accesses interrelationships and intertwinings in the moments when he is most intensely at work and in close contact with neighbors, animals, and the physical world. These moments are accessed in instants of intense reflection and made available by their retelling. Telling stories is one of the ways the community of Richardson is sustained. Haines incorporates into his own narrative long winter hours around the roadhouse table telling stories and drinking coffee spiked with rum: "A chronicle of the wise, the foolish, and the lucky—it will be resumed one evening when we are here again, to renew the playful innocence of an early day when men could stand in wonder at a beast, to marvel at a world abundant with things that walked and flew and swam and seemed possessed of understanding, to speak at times almost like men themselves" (*Stars* 39). The act of renewal includes a retrieval of an instructive past.

Wonder grows from a perception of shared human existence as it interrelates with the landscape and the rest of creation. Through sharing his experience of radical proximity to the physical and storied land, Haines involves readers in a shared ecological perception of community and world in which the cultural and the natural emerge as inseparable, ecotonal.

In the preface to *The Stars, the Snow, the Fire*, Haines writes: "In reliving parts of the narrative, I seem to have wandered through a number of historical periods, geological epochs, and states of mind, always returning to a source, a country that is both specific and ideal" (ix). The country is ideal in that it can provide him with a potential way of perceiving the world and his relation to it. It is also a source of the creative imagination. It is specific in that it is the physical place in which Haines lived and worked with the full intensity of experience, construed in Deweyan terms as an interpenetration of creature and environment. In the most forthright language, Haines sees the ecology of his watershed as a living entity: "Like all young rivers, the Tanana does strange and unpredictable things" (*Stars* 102).[3] The river has agency, it "does" things, and the landscape around Richardson participates in Haines's experience. Haines felt that to find a better way to perceive his relationship to the physical world and to his art, he had to go to ground in a place that was still largely wild:

> Why I chose that particular place rather than another probably can't be answered completely. I might have gone elsewhere and become a very different poet and person. But there was, most likely, no other region where I might have had that original experience of the North American wilderness. Unlike other "wilderness" areas, Alaska in those days seemed open-ended. I could walk north from my homestead at Richardson all the way to the Arctic Ocean and never cross a road nor encounter a village. This kind of freedom may no longer be available, but at that time it gave to the country a limitlessness and mystery hard to find now on this planet. ("Writer" 367).

He chose a way of living and writing that he saw as productive of a viable human participation in the ecological community. His manner of writing cannot be separated from his way of living, and the reference to open-endedness suggests an ongoing experience, left off in Alaska and picked up in this book.

Haines never encourages anyone to literally follow him to the Alaskan woods to write essays and poems. However, the book itself opens up the possibility for another type of experience for the reader, one that has consequences for how we perceive the world around us, which in turn contributes to how we understand ourselves as agents in the world. We experience the narrating self in this book as relational and begin to understand the self only in relation to text, reader, winter, and the ecology of the Tanana watershed. A complex level of narration in *The Stars, the Snow, the Fire* enacts this kind of textual agency. In the opening essay, "Snow," Haines meditates on ways of reading the snow, which seems perfectly natural for a person living off the land in subarctic Alaska. For one who traps animals there it is a practical skill acquired over time.

He also imagines a subarctic writer in this position: "I have imagined a man who might live as the coldest scholar on earth, who followed each clue in the snow, writing a book as he went. It would be the history of snow, the book of winter. A thousand-year text to be read by a people hunting these hills in a distant time" (5). Obviously, the trope of nature as a book is not quite as moribund as St. Armand thought in his discussion of Muir. The stubborn trope keeps surfacing, and for Haines it has become a metaphor for the practical skill needed for both a writer and a hunter in the northern landscape. However, Haines voices caution here lest we confuse him for the figure. When asked about his identification with this narrative construct, he replies:

> What I had in mind was more like a rhetorical, or metaphorical figure. I liked the *idea* of it, you see, and this followed more or less naturally from thinking about snow and the life I once lived in the snow. If you follow that thought, there is a close relation between reading the snow—reading signs in the snow—and reading words on a page. In fact, the relation between these is at the heart of much that I have written and talked about in one place or another.
>
> Anyway, think about it: "a man who might live as the coldest scholar on earth." There's a lot of resonance in that figure. But don't confuse that, literally, with me. Though I may have begun with myself, once I invented that figure it was no longer me; it became a thing with its own life. (Hedin 72)

Haines creates an explicit link between the scholar reading the environment —"signs in the snow"—and the audience reading this book—"words on a page." Additionally, the narrating figure, the coldest scholar on earth, is able to present the experience of living in the North in a way that was not necessarily available to the person actually living it at the time. This act of reflection and articulation grows from the desire to intertwine with other subjectivities, including those of the reader and the environment. By linking the cultural artifact (the book) to the natural phenomenon (the snow), the icy scholar shows forward the ability of the artist to form experience in a way that leverages open a space from which to criticize the destructive aspects of the culture in which most of us find ourselves embedded. By identification with this narrator, the reader also participates in the critique by experiencing how agency is enacted in the production of a work of art engaged with the natural world.[4]

Agency is foregrounded by Haines's attempt to present experience in an ecological way that includes human beings, work, and artistic endeavor in natural ecosystems. Haines sets his characters in relationships to the physical world in a way that makes their agency unmistakable: "I suppose that what you see of this in my own writing owes much to my having lived for so long in circumstances that more or less compelled me to see human activity on a background of such elemental scale that even today, with all the changes that have occurred, earth and the weather tend to overshadow nearly everything else. Yet it takes very little human interference to change things: one word spoken in an absolute silence, one shout, one rifle shot, means more, changes more, than a thousand people marching in a city where noise is the first element" (Hedin 72–73). Attentiveness embedded in silence prompts our awareness that each and every action has ramifications in the world, that "before knowledge, there was wisdom, grounded in the shadows of a dimly lit age" (*Stars* 157). Silence establishes a context for all that happens in *The Stars, the Snow, the Fire*. Haines continually functions in silence. In a discussion of Wendell Berry, another writer who prizes silence, Randall Roorda comments that in a raucous world, "Instead, silence . . . with its cognate 'quiet,' is to Berry the element and agent of learning in nonhuman wilderness" (113). Remembering after twenty-five years the evening he first heard Haines read his poetry, Berry himself writes: "Mr. Haines's poems, as I heard them that evening, told that they were the work of a mind that had taught itself to be quiet for a long time. His lines were qualified unremittingly by a silence that they came from and were going toward, that they for a moment broke" (25).

For Haines, speech is always speech about the silent world. Even the silence of a tiny animal like a bat can heighten an awareness of being that results in the articulation of silence: "And to think, from this long vista of empty light and deepening shade, that so small and refined a creature could fill an uncertain niche in the world; and that its absence would leave not just a momentary gap in nature, but a lack in one's own existence, one less possibility of being" (*Stars* 159). The self seems reduced by this loss, but Richard Poirier can help shed some light on how this reduction could be seen in a positive light: "Why may not reduction be associated in and of itself with exploration and gain? Part of some search for a world prior to the human presence and subsequent to it?" (213).[5] It can, but traditional humanistic tendencies to assert the ascendancy of the human self above all other possibilities often block any exploration of a silent world. This "momentary gap" is transformational in that it destabilizes the self, and "it is possible to confer value on moments of transformation or dissolution without looking ahead toward a narrative of fulfillment. The moment is endowed with something as vague as wonder or beauty, empty of the desire to translate these into knowledge" (Poirier 202). Wonder and beauty are sensed glimpses of "Empty light," "deepening shade," "absence," "gap," "lack," and a pause of uncertainty that all suggest silence without explicitly naming or knowing it.

Seen in such a long vista of light and silence, both human and animal, right down to the smallest individual creature, have intricate meaning in the larger context of "being," which in Haines is always tied to experience. As Dewey says, "Experience occurs continuously, because the interaction of live creature and environing conditions is involved in the very process of living" (*AE* 42). Haines's "being" and Dewey's "living" seem to imply, then, the interrelationship and intertwining of all things. "Being" seems to mean ecological possibility illustrated in the ecotonal nature of the relationship of narrator to bat, of writer to text and reader, and of all four to the physical world. And it is represented in Haines by articulation of silence, of a world that subtends all forms of earthly existence and human expression and that participates in *The Stars, the Snow, the Fire:* "only the wind and the distance, the silence of a vast, creatureless earth" (160). Poirier argues that readers and critics remain largely incapable of "looking at a landscape from which the familiar human presence has been banished and of enjoying this vista without thinking of deprivation" (203). On the contrary, in Alaska, Haines was intimately familiar with a landscape of diminished human presence, and he prizes that closeness to the nonhuman or prehuman world. He works to counter the notion of deprivation that Poirier sees rampant in our culture by connecting "being" with this prior and subsequent landscape. His attempt to represent "being" through silence is an ecological act that has the potential for far-reaching change in cultural domains once human beings accept that the physical environment is part of their physical and imaginative well-being, that it is part of their selves, though the human self is not necessary to its existence and continuance.

Haines grounds his essays on ecological "being" in *The Stars, the Snow, the Fire* on a period of years that were experienced in a very real place but which, of course, have been altered by memory:

> It is true also that certain experiences, states of mind, and ways of life, cannot be willed back. That intuitive relation to the world we shared with animals, with everything that exists, once outgrown, rarely returns in all its convincing power. Observation, studies in the field, no matter how acute and exhaustive, cannot replace it, for the experience cannot be reduced to abstractions, formulas, and explanations. It is rank, it smells of blood and killed meat, is compounded of fear, of danger and delight in unequal measure. To the extent that it can even be called "experience" and not by some other, forgotten name, it requires a surrender few of us now are willing to make. Yet, in the brief clarity and intensity of an encounter with nature, in the act of love, and (since we are concerned with a book) in the recalling and retelling of a few elemental episodes, certain key moments in that experience can be regained. On these depends the one vitality of life without which no art, no spiritual definition, no true relation to the world is possible. (x)

In this extraordinary passage are condensed the ingredients for a "true relation" to—or "an honorable passage through"—the world. It is helpful simply to list them: "intuitive relation," "experience," "fear," "danger," "delight," "surrender," "clarity," "intensity," "love," "recalling," "retelling," "regained." In a nutshell, these words could be used as descriptors for a state of consciousness requisite for an ecological perception of the world. Haines suggests that without the moments of insight that meld these responses, no art is possible, so for him both being and art depend upon this "true relation to the world." *The Stars, the Snow, the Fire* seeks to regain these aspects of experience that somehow exist beneath language but can also be partially regained by language. Because, as Haines says, "the experience cannot be reduced to abstractions, formulas, and explanations," it is expressed by moments embedded in silences that are the essence of his experience in Alaska.

Essence need not be seen as static; it is not a foundation, but consists of the whole of experience or is simply impossible to pin down completely because, as Dewey reminds us, it "emerges from the various meanings which vary with varying conditions and transitory intents" (*EN* 144). These essential moments, or points of clarity, contribute to a "safe and honorable passage through the world." Access to them does not require an Alaskan wilderness for everyone. Access depends, to take the title of a Haines essay, "On a Certain Attention to the World," which results in "delight, that sense of delight in discovery that renews everything and keeps the world fresh. Without it, poetry dies, art dies; the heart and the spirit dies, and in the end we die" (*Fables* 126). And we can find delight in a crack in a sidewalk or in a daffodil if we pay attention.

In *The Stars, the Snow, the Fire*, the shifty essence of Haines's experience resides primarily in work.[6] These essays on living in the North begin with a recollection of the skills of trapping, but three-quarters of the way through the book the essay "Mudding Up" focuses on work. "Mudding Up" is central to this discussion for several reasons. First, it recounts an incident from Haines's initial year in Alaska, and from the beginning, the physical world participates in the work Haines performs. Second, it highlights how shared work can be one of the principal steps in the formation of a community. Third, work as a retrieval of ways of life and skills no longer widely known leads to deep satisfaction. Work ultimately becomes linked to the action of writing these experiences, making them available for a wide readership: "The experiences both of reading and of the writing it creates are more real, more present to consciousness, than are any prior circumstances that might have given rise to them" (Poirier 201). The reading of the past bears on the present.

As is often the case in this volume, the essay opens in profound silence: "A clear afternoon in mid-September 1947. You will know the kind of day I mean: a warm, tawny light over the hills, the sky is flawless in its clarity, and already in the windless air leaves drift down from the birches and aspens" (104). The second-person address to the reader and the casual insistence that the reader will know exactly what the narrator describes ease the reader into the essay. All is silent in the subdued light of Haines's painterly prose, and the quality of the day is described as "A day when a shout, an axe stroke, or the single cry of a raven, rings clear and remote. The humming of a few late wasps searching the browned marigolds, the drone of a bumblebee, surging and resting, fills the hollow of all creation" (104). Human activity is clarified against a backdrop of silence—"the hollow of all creation." In other words, actions are performed in the context of an elemental world. Against that world, the slightest human task is foregrounded, and human agency is not, as one would expect, overshadowed by the larger world, but isolated in a way by Haines's prose, which makes all human action as clear as the day, as resonant as an axe stroke. So, human work—an axe stroke—participates in the larger environment.

Haines and his friend Fred Allison, an old-timer at Richardson, are mudding up the log walls of an old stable so that Allison can use it for a chicken coop. The older man shows a level of expertise attained through long practice: "Allison is on the ground below, mixing the peaty soil with water into a square-sided five-gallon tin opened lengthwise to make a shallow trough. With a practiced stroke he stirs the brown mud-mortar back and forth with a stump of a hoe" (105). Haines stands above on a ladder and forces the mud into the chinks between the logs, and in the mantra of the work, he begins to lose himself in thought. He returns in his mind to the war and to the "cold, slant light" of New York City, and then:

> I came back to this present moment, and to the thing before me, to a simple task of repair I could take pleasure in learning to do. And to whatever was a part of that: the glint of mica in the sandy clay soil, the ragged peat moss sticking in the log seams; to nothing more than the split and knotted, weather-greyed wood of the wall in front of me. I saw my hand on the trowel, and the wet brown mud heaped on the board I held; and below me by the stable wall, Allison's broad and ruddy face in a stray shaft of light. I heard the soft scrape and slop of the water and the earth he stirred back and forth in the gleaming tin trough. (106)

Again, attention is heightened, and Haines's focus narrows down to a near view of the very glints of mica in the mud. Haines and Allison are linked through their work to the materials they use to repair the wall described in these terms: "mica," "clay," "soil," "peat moss," "log," and "wood"; Allison stands in a "shaft of light," linking him to Haines in the slant light of his reminiscence. All these components of work are basic elements of the physical world. Haines seals the passage in the final sentence by shifting the register of his prose and transforming the materials from specific things to "the water and the earth." The water and the earth and the light are the comprehensive elements that support all life, and, important here, they are also "whatever was a part of that," "that," of course, signifying the "simple task of repair" that Haines enjoyed learning. That task of repair as an act of work incorporates the physical world and links Haines and the reader to that world. The articulation of this incorporation and linkage in compelling prose, thus making it available across subjectivities, is itself an act of repair.

And repair work leads to reflection: "As concentrated as I was on the moment, savoring every detail, did I sense that this quiet, rural world of Richardson, with its few surviving people and its old-fashioned implements, remote and settled on a stretch of gravel road, was vanishing even as I came to know it? It may have been that in some hidden part of myself I knew this, knew that I in some way was a part of the change taking place. But for now there was this moment, this day, and the promise of others to come: a vague, but tangible dream realizing itself in the cool, diminishing light of a first fall" (106). Using a complex narrative strategy, Haines forefronts the fact that this is a retelling, and this retelling is centrally important to the logic of a pragmatist ecology. In the passage is the juncture of the tangible and the intangible, the dream and the "diminishing light of a first fall." Haines manipulates our gaze so that we perceive that which is just beyond our focus, just outside the spare light of autumn. The language is ethereal and lyrical, yet solidly linked to the concrete, "a stretch of gravel road," for example. The linkage of ethereal and concrete has its analog in Dewey's work as well, in the linkage of realism and idealism. In "The Development of American Pragmatism," Dewey insists that "Logic, therefore, leads to a realistic metaphysics in as far as it accepts things and events for what they are independently of thought and to an idealistic metaphysics in so far as it contends that thought gives birth to distinctive acts which modify future facts and events in such a way as to render them more reasonable, that is to say, more adequate to the ends which we propose for ourselves" ("Pragmatism" 18). The realism of Haines's Richardson becomes interrelated with the idealism of his dream in the chill of the falling autumn light.

What becomes evident in Haines's work is its attentiveness to the real, physical world as it looks at the same time toward the potentiality of the future. It sees the actual in the light of the possible. Through its focus on a way of living close to the world, along with the world, his work suggests that we can begin to think about a better way to inhabit our places. Pragmatist ecology proposes that we live in a manner that does justice to the physical world on which we depend for our very lives; that we do justice to the community in which we live, which in its turn means that we care for the least of us in the equation; and that we begin in a very democratic sense to understand that each and every one of us can contribute to a democratic community. Without this belief in the ability of each to contribute, the dream of democracy will continue to wither. We can contribute through our work, and work can provide a portal through which we can perceive our relations to physical and cultural ecologies.

When Haines begins to ponder what he knew as he worked on the log structure, he obviously speaks from the moment of composition, and he uses a key phrase that can help us better envision how the past can work on the present and the future. When he writes "But for now there was this moment," he creates a complex nexus of time by insisting that "for now," for the present time of writing, there was "this moment" from the past. He intentionally conflates the past with the present, dragging the past experience into the writing experience. This is further supported by the lines following: "Time passes. The work goes quietly; my mind drifts on ahead by days and weeks. It is afternoon once more; the veiled sun lies much lower in the south, spending its cold, grey light over the river and the fields. A light snow now lies on the ground, no more than an inch over the frost-shattered stubble of grass" (106–7). The work on the building and the work of writing are contained in the same phrase, "the work goes quietly," and "quietly" sets the work in the context of a silent world. Time is passing as he sits writing, and the description of passing time is accomplished by reference to the position of the sun in the sky and to the snow on the ground. In the continuity of experience, time, weather, work, physical environment, and writing come together in the present(s) of this short section of "Mudding Up."

Haines surely is not simply writing for his present moment, unless, perhaps, we think of the present in Deweyan terms as "the dynamically insistent occasion for establishing continuity or growth of meaning. Present experience stands for that whole complexity which establishes the human project as such" (Alexander, *Horizon* 269). The present moment is of central importance because it represents a nexus of past and future, both of which bear on the present. Haines seems convinced that his recollection and retrieval of a past way of life can affect a present and a future readership, "And still that lost being pursues us, no matter how remote and abstract our sensibilities have become" (*Stars* 158). In his meditation *Nostalgia*, Ralph Harper writes that nostalgia "constantly reminds, through the pang of what is lost, what might be found" (87). The past is not something to wallow in, but a thing to be brought to bear on the future. It is a critical tool. In a similar context, Raymond Williams examines the relationship between the country and the city. By simply dismissing nostalgic feelings for the past as yearnings for vanished or imaginary rural idylls, he insists, we relinquish a potent critical power mobilized by looking at the present through the glass of the past (297). Haines's evocation of a past way of life bears critically on the present. In a recent newspaper article I came upon the phrase "nostalgia doesn't pay the bills," and this is just one of the ways in which the lenses of the past are easily and thoughtlessly shattered and discarded in our culture ("Board"). The primacy of bottom-line thinking in our culture can and often does obliterate our link to the built, cultural, and natural past. The loss of the past can be extremely dangerous. Harper, writing in 1966, resonates deeply with contemporary U.S. culture as the mythology of Vietnam becomes again a blurring lens:

> Men lose their sense of the independence of history when they keep windows to the past open only by habit or by propaganda, instead of by study and wonderment. They do not understand that the journey to the past is only a detour to the future, prescribed for human voyagers because man does not carry all his absolutes within him. Man is always mending his roads, tearing up stretches he has worn through, using the back roads as substitutes until he can lay down new road beds. Tyrants understand that this is the process of freedom which obstructs the achievement of simplified solutions. They know that they can control men most completely when they control their memories. The only return they let men envisage—and they permit this natural tendency to function so far—is a return to a mythical past which their propagandized present or future imitates. (98–99)

Recollection engages the future. As in "Mudding Up," the work of repair initiates perception of the relation between past and present. Mending is activated through study and wonder—the art of knowing—not through the propagandized past that U.S. culture continually and passively allows itself to be spoon-fed.

Of course, nostalgic links risk descending into the overly sentimental, and according to Williams, "often it is then converted into illusory ideas of the rural past" (297). Haines's relationship to his rural past, on the other hand, is anything but illusory. His essays are indeed deeply elegiac. But the ties Haines presents are ones that have come about, again, from his living in a radical proximity to the land in the middle of the twentieth century. In Dewey's thought, "The union of past and future with the present manifest in every awareness of meanings is a mystery only when consciousness is gratuitously divided from nature, and when nature is denied temporal and historic quality. When consciousness is connected with nature, the mystery becomes a luminous revelation of the operative interpenetration in nature of the efficient and the fulfilling" (*EN* 265). Haines connects consciousness to nature through images and scenes of work, and as we have seen, past and future are interrelated with the present; real and ideal coexist as do the efficient and the fulfilling. In this collection of essays, Haines evokes a lifeway—and its accompanying form of consciousness—that is in the process of disappearing and puts readers into contact with the values he perceives in such an existence.[7] The essays are brutal, lovely, and sad. And, even without overt condemnation, they point to our own dangerous, often self-willed alienation from our physical environment. For Haines, "The land lives in its people" (*Stars* 146).

It is worth thinking a bit more about nostalgia at this point. Williams wryly comments that "Nostalgia, it can be said, is universal and persistent; only other men's nostalgias offend" (12). But he also argues that "A memory of childhood can be said, persuasively, to have some permanent significance" (12). So too, then, must an adult's recollection of a meaningful time in his or her life recalled from a perspective of thoughtful distance possess permanent significance. I suggest we complicate our notions of nostalgia. As in the example from the newspaper article, "nostalgia" normally evokes thoughtless dismissal of values discovered from the past. Certainly, nostalgia can be a weak-minded yearning for things that are gone, can be sitting around watching *Happy Days* reruns or pining away for a Currier and Ives rural America that never existed. Along these lines, Haines writes, "the original beauty, the instinctive awe and mystery, become in time a matter for nostalgia and of scenic views" (*Fables* 126). In this case, Haines seems to use nostalgia in a rather careless way, though he later credits nostalgia with great power. Nostalgia deserves more attention as a powerful critical tool, I think. It is for Harper a kind of involuntary conscience—it is positive—an impetus to reconstruct (26). He writes: "in nostalgia one smells and tastes, one responds from the darkest corners of oneself, as a renewed whole, to some reality one loves, a person or a place or even an idea. No longer is there any excuse for waiting: nostalgia is regenerative and requires the starting of life all over again" (28). The recovery of the past, then, is a tool that inspires action. Edward Abbey has also weighed in on nostalgia: "Suppose we say that wilderness invokes nostalgia, a justified not merely sentimental nostalgia for the lost America our forefathers knew. The word suggests the past and the unknown, the womb of earth from which we all emerged. It means something lost and something still present, something remote and at the same time intimate, something buried in our blood and nerves, something beyond us and without limit. Romance—but not to be dismissed on that account. The romantic view, while not the whole of truth, is a necessary part of the whole truth" (166–67). Again, there is an interrelation of past and present that can be brought to bear on the future, on a better way of living in the world.

Haines at other times clearly acknowledges the power of nostalgia, and it seems clear that a past portion of a life lived in close connection to the land can perform a powerful critique of current ways of perceiving our lives in this culture at this point in time. Nostalgia can be understood as possessing critical leverage. *The Stars, the Snow, the Fire* itself grows from a powerful nostalgia:

> I realized a long time ago, when I had finished *Winter News,* that there were things that needed to be said, that at the time I could see no way of getting at in poems. Once, perhaps, in my younger days, I might have tried to write it out in a narrative poem, or as a dramatic dialogue, but I didn't do it that way. I was encouraged once by Donald Hall, back in the sixties, to try and write it out in the form of the personal essay. I didn't know how, and I was for some reason afraid to try. I was too close to it. I didn't begin to write about it until some time in the seventies. It was not until I was living in the north of England, in Yorkshire, in the spring of 1977, with the onset of spring there, and with it an acute nostalgia for the north, that the ideas began to come. By that time, I had enough distance. (Bezner, "Interview" 15)

Since the book was not published until 1989, this feeling of "acute nostalgia" led Haines to a profound level of thinking about his own experience and about the ramifications of that experience for the culture as a whole.

Haines lets us know from the beginning of the book that "This journey in and out of time cannot be adequately expressed by any sum of calendar years. In the sense in which I write, there is no progress, no destination, for the essence of things has already been known, the real place arrived at long ago" (*Stars* ix), although their essences need not be understood as stable. However, the book is not really a movement in and out of time. As the passages from "Mudding Up" clearly show, what Haines describes is more like a continuum of time, where events from the past inform both the present and the future. The world is not a stage on which we observe life—past, present, and future—going on. In fact, as Dewey writes in *Experience and Nature*, "In creative production, the external and physical world is more than a mere means or external condition of perceptions, ideas and emotions: it is subject matter and sustainer of conscious activity: and thereby exhibits, so that he who runs may read, the fact that consciousness is not a separate realm of being, but is the manifest quality of existence when nature is most free and most active" (293–94). The porous boundary between consciousness and the physical world is the locus of possibility, the field of potentiality where all things have the possibility to coexist, contribute, and participate. In other words, the nexus of consciousness and natural world as a field of compossibility is a site of ecological and evolutionary potential—an enormous ecotone. It is, then, for us to act upon the possibilities inherent in living in a world conceived in this way.

Work is Haines's access to this environing world, and work can also be seen as an overarching metaphor for what needs to be done in the attempt to construct communities that exist in a viable relationship to the environing world. In his prose, work opens up these possibilities, allows us to glimpse what lies mostly on the fringes of our vision. Haines opens the essay "Other Days" in much the same way he opens "Mudding Up," with a mention of the month. He sits on the porch of his cabin mending snares as he talks about cutting wood, about a pile of sawdust on the ground. With no warning we find ourselves inside the cabin in the evening, and "Out the window, in the southwest, a cloudy light fades slowly over the mountains. The river channel at the foot of the hill is frozen, but downriver I see a dark streak in the snow: open water" (85). The view from around the cabin is not particularly spectacular, and Haines focuses primarily on the more mundane tasks of existence in the woods.

Then, with the river—the agent of change—as cue, he writes, "The land changes slowly in a thousand years. The river has shifted from one side of the valley to another, worn its bed deeper in the sediment and rock" (85). So far we are confronted with the silence of piles of sawdust, ice, water, and the deep silence of geological time. Again, in these connections of work and world Haines functions at his subtle best. As he sits mending snares, mundane tasks are linked to the geomorphological work the river itself performs on the land. The work of the human and the work of the river are aligned, and the smallest human endeavor in the largest natural landscape becomes on a scale with geological time. Experience is aeonian.

This is essential to what Haines wishes to say in his book: that every one of our actions is crucially important to the future of our particular places and communities, to the future of the culture, and to the future of the natural landscape. By the end of the essay, he can write that "In this immensity of silence and solitude, my childhood seems as distant as the age of mastodons and sloths; yet it is alive in me and in this life I have chosen to live. I am here and nowhere else" (87–88). In this forthright statement, Haines articulates a fact that seems so simple that we are likely to read over it—"I am here and nowhere else." Many of us, however, would have great difficulty honestly claiming this. For instance, if I decide to search for a job, and I look at possible places of employment on my computer and fire off e-mails, I am everywhere. I am also nowhere—not at all rooted in where I find myself at the present moment. As we allow our lives to become increasingly homogenized, the particularity of our local places begins to vanish into a Wal-Mart world. Dewey argues in *The Public and Its Problems* that "Democracy must begin at home, and its home is the neighborly community"; he goes on, "The local is the ultimate universal, and as near an absolute as exists" (368, 369). The local, neighborly communities upon which democracy depends are being quickly eroded. For many thinkers, the decay of these communities is also tightly knit with degradation of the environment. In a distinction drawn by David Orr, most of us are residents, not inhabitants: "The resident is a temporary and rootless occupant who mostly needs to know where the banks and stores are in order to plug in. The inhabitant and a particular habitat cannot be separated without doing violence to both. The sum total of violence wrought by people who do not know who they are because they do not know where they are is the global environmental crisis" (102). Work such as *The Stars, the Snow, the Fire* can help us repair our sense of place. Haines knew where he was, and I suspect that leaving Alaska did him some violence. He insists that his childhood is analogous to his region's. He inhabits it as the mastodons did. His focus out from the smallest details to embrace the everyday experience lived in an ecological relationship to a particular place voices a powerful critique of the way many of us live our unplaced lives as residents at this point in history.

A cabin is a very particular place, and Haines often relies on a cabin or building to set a scene. The cabin is a powerful figure in his writing—cabins are one of the end results of his work. Since *Walden* was published, any mention of a cabin in the woods immediately evokes both nature and book, and Roorda writes interestingly about the cabin as a figure in ecological writing: "The cabin is the scene where this identity of the subject and site is ideally transacted, with the window emblematic of the sought-after simultaneity of 'living' and 'writing.' In practical terms, neither the elements of nature nor the machinations of writing need be diminished in their contact across this boundary: the paper stays dry and the trees stay wild" (165). Haines takes the cabin site a step further. He uses the cabin, and the log stable in "Mudding Up," to link not only subject and site but to link work in general to the economy of the natural world. The cabin's porch and roof shelter him as he works. When the above scene opens he is making snares and cutting wood, and something is cooking on the stove. The nexus of these activities, the cabin, shelters his work, and work is the middle term connecting living—or being—to writing, and the transaction is sited in the cabin or performed on the log walls of the stable.

Haines also read, wrote, and labored there, so the cabin is both a physical and mental hub. Trails run from the cabin out into the world and back into the mind. The cabins he built are vitally meaningful to him in the present, and they represent a point from which he moves off into a subsequent world void of the human self: "I may not always be here in these woods. The trails I have made will last a long time; this cabin will stand twenty years at least before it falls. I can imagine a greater silence, a deeper shadow where I am standing, but what I have loved will always be here" (76). Part of what makes these structures so important is not only what they house but also what surrounds them: the boreal forest, the river, geological time—silence. They are a kind of portal through which Haines imagines a movement from egocentric toward ecocentric existence, imagines the self engaged with the enveloping silence, participating in the environing world. The built structures in the prose function as mediational nodes, understood as a site of communication and participation between the subject and the environing world. Haines senses, as does Dewey, that our experience of the physical world is nearly always mediated: "But only a twisted and aborted logic can hold that because something is mediated, it cannot, therefore, be immediately experienced. The reverse is the case. We cannot grasp any idea, any organ of mediation, we cannot possess it in its full force, until we have felt and sensed it, as much so as if it were an odor or a color" (*AE* 125). These mediational nodes are part of the language of love and work and silence.

And that silence always lurks just outside of the cabin: "Before going back up to the cabin I stand for a moment and take in the cold landscape around me. The sun has long gone, light on the hills is deepening, the gold and rose gone to a deeper blue. The cold, still forest, the slim, black spruce, the willows and few gnarled birches are slowly absorbed in the darkness. I stand here in complete silence and solitude, as alone on the ice of this small pond as I would be on the ice cap of Greenland" (69). Silence surrounds everything in this icy corner of North America, and it extends beyond. In fact, by linking two extreme points of North America, Alaska and Greenland, Haines can be said to include everything in between, and amid the din of our technological world, "silence, or quiet, whether in one's working space or neighborhood, or in the woods, can be of enormous significance; and I mean that silence in which it is possible to hear many things disregarded" (qtd. in Coffin 54). This silence, of course, allows one to hear the world. Harper writes: "In silence men love, pray, listen, compose, paint, write, think, suffer. These experiences are all occasions of giving and receiving, of some encounter with forces that are inexhaustible and independent of us. These are easily distinguishable from our routines and possessiveness as silence is distinct from noise" (115). For Haines and for a pragmatist ecology generally, one of these encountered forces is the physical world. Silence allows one to be attentive and to take part in the ecology of one's surroundings.

Participation in an ecology in this sense demands silence in order to let the other speak, and this idea entails an ethical consideration for the natural world that seems to fall in line with both Roorda's and Attridge's arguments for an ethical consideration of the other. I must be willing to suspend habits and familiar ways of thinking in order to be ethically responsible for the other. This responsibility demands quiet, and Roorda also suggests that quiet is doubly significant: one can be quiet or one can be in quiet (134). This double relationship insists upon a deep engagement with the external world and an attention to the world that recognizes it as a participant in experience. There is an intertwining and reciprocity between self and other, human agent and physical world.

And, of course, the physical world in the Far North manifests itself not only in silence but also in snow. The cabins are built for the snow, and windows become a metaphor for the reciprocal insertion and intertwining of subject and object, of seer and seen: "When we built this cabin, I set the windows low in the walls so that we could look out easily while sitting. That is the way of most old cabins in the woods, where windows must be small and we often sit for hours in the winter, watching the snow" (69). Naturally, there would be long periods of time when one would do nothing but watch the silence of the snow, and this, of course, brings us back around to "the coldest scholar on earth." Throughout the book are references to reading the snow, for instance, "I walk, watching the snow, reading what is written there, the history of the tribes of mice and voles. Of grouse and weasel, of redpoll and chickadee, hunter and prey" (65). Recollection, reading, and precise ecological knowledge are embedded in the snow: "If I consider it now, with many details forgotten, . . . what I return to is the deep wonder of it. How it was to go out in the great cold of a January morning, reading the snow, searching in the strongly slanted shadows for what I wanted to see. And there were books to be read there, life histories followed sometimes to their end: a bit of fur matted in the stained ice, the imprint of an owl's wing in the snow" (28). There is deep wonder at the interrelationship of the creative intelligence, the world, and the text. The history of Haines's time homesteading in Alaska is recorded in his writing. The histories of the animals that inhabit Alaska are written in snow: "Blood into ice, and fur / into matted frost— / this is the way of winter, / on earth as in heaven" (*New Poems* 9). These inscriptions are in books, history, poems, on snow, in the seasons, and as the line of poetry taken from the Paternoster implies, in prayers. This suggests that the landscape and the animals who live in it are as much a part of our culture as are our computers, movies, and automobiles—a far deeper part, in fact. Our culture is driven by writing, and Haines, of course, knows this. But he also knows that these creatures and these landscapes so different from us, these animals whose writing we cannot read and these silent places to which he has lived in radical proximity, also deserve to have their books and histories written and read. Haines attempts to make these creatures and places legible to us by forging the link between the book and the snow. Through their textualization, he brings them back into the world as participants in our current experience.

William Cronon makes a similar link in a *New York Times* editorial about drilling for oil in the Arctic National Wildlife Refuge. He makes the interesting claim that the refuge is, contrary to the usual rhetoric, neither barren nor remote. The land is inhabited and considered sacred by the Gwich'in people, and Cronon makes a statement that cuts to the heart of our misconceptions about remote places:

> Among the 180 bird species that use the refuge is the tundra swan, once more familiarly known as the whistling swan. After raising their young, these birds migrate thousands of miles across the continent to their winter homes along the Atlantic Coast from North Carolina to Maryland. From the perspective of a tundra swan, Washington, D.C., and the homeland of the Gwich'in are part of a single ecosystem.
>
> If migrating birds remind us that the neighborhoods where we live are in fact linked to the refuge, then we should also remember that how we live is what puts the refuge at risk. The ways we drive our cars, heat our homes and otherwise consume oil are the biggest single threat it faces.

These animals and their habitats are all intertwined with our lives, and Cronon forces the critical eye both inward and outward. Haines's and Cronon's words, the tracks written in the snow, the migration paths of these birds inscribed on the sky, and their whistle should remind us of how our paths are, more often than not, intersections.

To someone living in the woods, of course, paths have a very practical function. If one runs trap lines, as Haines did, one must blaze trails through the woods, establish caches, build a few cabins along the way. Paths, roads, and trails all have a strong metaphoric resonance in our literary tradition, and they are important to Haines's retelling of his time in the North. After all, he blazed them, and they were the initial key to his economic survival and integral to his very physical survival: "I put a lot of care into making those trails, and I take some pride in knowing that most of them are still there, sound and true. A trail through the woods is made for a purpose; and if it is important enough it is worth the time spent to do it well—or so I thought as I sighted a way through the birches to the next rise, and checked behind me to see that the way was clear and the grade as easy as I could make it" (*Stars* 14). A trail is both work and ecological perception. Roorda points out that whereas roads and airplanes move over the landscape, a path proceeds through it (118). To walk through the landscape only as fast as one's feet can travel is to retain "the contact of a path" (126). To pay close attention while moving through the landscape on a particular path is an attempt to engage the landscape in an ecological way, a way that is gained by slowly earned familiarity, which can lead to knowledge that one participates with the place one has come to know well. Different kinds of trails were cut, however, during the time Haines was there; rights of way were cut through the woods for pipelines and power poles.

So Haines does not work in a pristine wilderness, if there is such a thing. The trails themselves begin to assume the life of the place, becoming far more complex than simple paths through the woods. By intersecting with older inscriptions on the land, the trails assume a historical significance as well as becoming integral parts of Haines's present world: "With daily seasonal use these trails became in their own essential way a part of the homestead, an extension of the yard" (*Stars* 14). The trails, then, allow Haines to proceed into the world. They are useful: "All things encountered along a trail might be of use—a dry snag for kindling, a dead birch for the bark that held it upright, a dry leaf floor under aspens for a crop of mushrooms in late summer" (14). The trails help Haines survive, providing him access not only to game but also to vegetables and fire. They aid in his education. Knowledge, for instance, that it is only bark holding up a punky birch tree could only through long stretches of time in the woods (or through reading this book).

Further, the trails begin to tell their own stories: "In no time at all the trails acquired their home legend of past kills and other memorable events—here a bear was feeding early in the summer, and there last fall a bull moose stripped a spruce sapling of its branches with his horns" (14). The trails now take on an extraordinary agency; they are capable of telling their own stories, and presumably they should be able to tell stories of the human beings who wander along them as well. First they locate Haines, through work, in a particular ecology: "The labor of it occupied me the better part of three years, but I have known and done few things more satisfying. I contemplated the map with a sure sense that I knew where I was in that far corner of North America" (15). Here it becomes quite clear not only that Haines's essay collection brings a way of life to bear on the present but also that his writing is meant to investigate our role on the North American continent.

Haines lived in cities and towns all over the United States, but in Alaska he learned his place in North America, and for him, to know one's place is to know the landscape intimately, to cut trails into it, to learn from it, and to allow it to tell its own stories, to be "along with" its silence. To believe in and to convey this quality of living along with as opposed to on the land and to suggest thereby that such a life may point to a better way to exist on this continent is a profoundly ethical act that extends ethical consideration to the landscape itself as a participant in human experience. He says, "I see my withdrawal into the wilderness as one more attempt to understand what it means to be here in this place, America, and in a true sense to settle myself in that place" (qtd. in Cooperman, "Interview" 113). What Haines sought to understand, I think, is what Barry Lopez, in a book he dedicates to Rachel Carson, claims we need to rediscover, namely, our role on the North American continent: "What we need is to discover the continent again. We need to see the land with a less acquisitive frame of mind. We need to sojourn in it again, to discover the lineaments of cooperation with it. We need to discover the difference between the kind of independence that is a desire to be responsible to no one but the self—the independence of the adolescent—and the assumption of responsibility in society, the independence of a people who no longer need to be supervised" (*Rediscovery* 49). Lopez too points out the key terms we need to articulate in an ethical stance toward the physical environment, and they are terms and themes that course strongly throughout Haines's entire oeuvre: "cooperation," "difference," and "responsibility."

The construction of trails also involves cooperation and responsibility and the ethical recognition of difference—not an exploration in the colonizing sense, but an exploration of the interrelationship of self and land. In this vein, Haines realizes that paths and trails lead not only toward the external but toward the internal as well; they provide physical and psychic sustenance: "The physical domain of the country had its counterpart in me. The trails I made led outward into the hills and swamps, but they led inward also. And from the study of things underfoot, and from reading and thinking, came a kind of exploration, myself and the land. In time the two became one in my mind. With the gathering force of an essential thing realizing itself out of early ground, I faced in myself a passionate and tenacious longing—to put away thought forever, and all the trouble it brings, all but the nearest desire, direct and searching" (*Stars* 17). Of course, Haines cannot outrun thinking, and he does not seem to want to for long. But he also insists upon studying through feeling, through contact with things "underfoot"; that is, by studying the ground he is walking on, and by studying his own process of walking, he attempts a kind of study that does not demand thought, but, suspending the ego, insists upon contact. He seems able to close down thought momentarily. Subject and object—Haines and the land—are recognized, as Dewey insists, as inseparable. Thought inevitably returns, and in an important gesture, Haines marks his sense of responsibility by setting limits. As soon as he considers accumulating more land, he has "big and indefinite dreams never realized, though I could imagine them down to the last detail—the camps I would build, the trails I would cut, the fall hunt coming on early near timberline" (17–18). They are never realized, because "for me finally there were limits. Things at home also had their claim on me—that other world of books, and of thoughts that went far beyond such immediate things as hunting and trapping, into a country of their own. I would stay here and make the best of what I had" (18). Thought and action have different limits, and this realization has consequences for how we live: our ideas, if put into action, must take the limits of the environment into consideration. Haines makes the kind of commitment that defines an ecological relationship to the world. He acknowledges that he has limits, that he cannot expand his domain indefinitely out into the world. Other things have claims to him. Not the least of these are his home place and the creative imagination. His perception of the relations between the two lead to directed action in the form of writing and sharing his experience.

Haines moves away from his trapping life for various reasons, especially once income begins to come in from other sources. Trapping helped initiate him into the ecology of Richardson and the surrounding drainages, and though trapping is a form of work that led him to an uncomfortable relationship to animals, that connection clarifies immediately that the human being is also an animal that kills to live, although most of us no longer participate directly in killing, but buy, for instance, our meat at a supermarket counter. For Haines, the killing involves regret. He becomes adept at trapping, killing, and skinning animals, but these acts always make him uneasy, as in the following painful-to-read description of the killing of a fox. After stunning the fox just as Allison had taught him, Haines realizes that "He would not stay that way for long, so I quickly knelt down in the snow. I seized the unconscious fox by his forelegs and drew him into my lap. Holding him there with one hand, I grasped his muzzle tightly in my other hand and twisted his head as far around as I could, until I felt the neck bone snap, and a sudden gush of blood came from his nostrils. A shudder ran through the slender, furred body, and then it was still" (11). The slow, detailed description enacts Haines's fascination with killing the animal, and it is indeed difficult to turn away from the description. But Haines is disturbed by his act: "I released him and got to my feet. I stood there, looking down at the soiled, limp form in the snow, appalled at what I had done. This is what trapping meant when all the romance was removed from it: a matter of deceit and steel set against hunger. But I had overcome my fear, and I felt something had been gained by that" (11). After awhile the adventure of trapping wears off for Haines, "but," as he says, "I was at home there, my mind bent away from humanity, to learn to think a little like the thing I was hunting" (28). This is another attempt by Haines to identify with the landscape and the animals in it. Throughout the book there is the tension between the need to kill for livelihood and food, on the one hand, and compassion for the lives taken, on the other: "Yet I cannot trap and kill without thought or emotion, and it may be that the killing wounds me also in some small but deadly way" (75). Perhaps this refusal to fully embrace a predatory way of life is itself a marker for self-imposed limits. There are limits to an identification with other life-forms, and the crossing of those limits performs a violence that causes Haines unease. Although we cannot wholly avoid this violence, and often go to great lengths of self-deception to convince ourselves that our continued living does not cause the ending of other lives, his setting limits in *The Stars, the Snow, and the Fire* represents, perhaps, a decision to assert his humanity in the face of his animality, thereby coming to a realization that, for him, it was better to let other lives be as far as this was possible. He participates respectfully as a human creature in a world rife with otherness.

In his experience in the North, Haines marks what was and what remains important for us, and for our art. It is also important that his claim to value in this experience is not prescriptive. He believes in a renewal of language and art, and for him this entails a going back to what he calls primary sources, which, as we have seen in Carson, lead to a concern with continuity. But approaches to sustaining this continuity vary with different people and artists:

> But I think it's a valid criticism of contemporary poetry that it lacks for the most part a sense of beyond the occasional and the immediately personal. For one reason or another, however it has come to

me, I seem always to have had an intuition of some greater, larger
design in existence, but it has taken me a long time to define this to
any extent. And it's certainly true that my going to Alaska to live as
I did during those years was an essential part of my self-education—
not that it was deliberate, or that I had conscious designs in doing
it. But to go *back,* and to live for a certain time, though I don't really
like the term, a kind of archetypal life—to live by hunting and gathering, by fire and fish and blood—it was that kind of thing that gave
me a sense of primary sources and of a certain continuity in human
life. But after all, Yeats and Eliot and other poets in very different
ways acquired something like this, so I don't view my own experience as a universal prescription: it just happens to be the way I had
to go in order to learn what I needed to know. (qtd. in Bezner, "Interview" 11).

Although, as Haines notes, his experience is particular to him, he un-
easily suggests that it could be a kind of model or archetype. When that
experience is transmitted in essay form, the experience of writing and
reading establishes continuity between Haines's past experience in the
North—his own quest for continuity—and the audience's experience of
rereading and renewing that experience. Sam Hamill writes that "in the
age of Exxon, all of our furnaces are burning, but a chill remains. John
Haines locates a solution only within each of us, a way which can be
found only through insight and contemplation" (181). This is not simply
self-contemplation. Contemplation must also be focused on the intersections between self and world, and Haines transmits a structure of feeling about Alaska that initiates the perception of a close and vital interaction of human and nonhuman, most often found just on the fringes of
our consciousness.

Haines is equally concerned with the community he found when he
went to Richardson and of which he became a contributing member. His
concept of an ecological world includes deep respect for the animals and
the landforms they and we inhabit, includes the deepest oceans and the
most remote glacier. He has repeatedly voiced the conviction that we degrade the natural world to our own degradation:

> We are unable to see nature for the inclusive force that it is. It is
> about us and within us, a necessary part of everything we are and
> can do, of everything we make, even the books we produce. It is in

the dust of our carpets, in the spiders and roaches that seek whatever household crack or crevice exists for them, and in the shifting foundations of our buildings. It is in the squalor of the streets and neighborhoods of our cities, where nature, shunned otherwise, returns in the form of random violence and deliberate criminality. It is in the decay of our political systems, and in the confusion and incoherence of our public discourse. ("Rise" 83)

Both the environment and democracy suffer. But communities of people living with an awareness of their environment and finding value in their everyday work are essential to Haines's work and to a reimagining of our currently destructive ways of living in the world. A large part—perhaps all—of Haines's effort in *The Stars, the Snow, the Fire* is to catch and hold a local community of people he loved in the web of the natural world and in the patterns of his prose:

> They are useful ghosts, these old inhabitants with their handworn implements, their settled lives. They tell us something of what we have been, and if we live long enough and well enough, what each of us may become: one more sign of our residence on earth, alive by reason of remembered love.
>
> I was lucky to have known them when I did, for they are no longer standing in their patched wool and mended cotton. In some way I have always accepted, they were my people, if the phrase now means anything, and the best of them I have loved with a deep appreciation that has never left me. They were friends and teachers, and I do not expect to see their kind again.
>
> When I think of them now, it is of something hugely tender and forgiving, akin to a healing thingness in the world that assures the soil of its grasses, the earth of its sun. (147)

For both Haines and Dewey, the work of healing our democracy and our environment begins in the rejuvenation of local communities. Haines knew, loved, and worked with the people in the Richardson community, and his work then and in this book does them and their work honor. His work allows them to remain useful, and provides their and our access to an otherwise silent world that is in itself capable of healing in its physicality—in its "thingness." Richard White condemns our tendency to seek in the natural world human absence. To insist upon dividing off natural and human is finally futile and destructive: "And if we do not come to terms with work, if we fail to pursue the implications of our labor and our bodies in the natural world, then we will return to patrolling the borders. We will turn public lands into public playgrounds; we will equate wild lands with rugged play; we will imagine nature as an escape, a place where we are born again. It will be a paradise where we leave work behind. Nature may turn out to look a lot like an organic Disneyland, except it will be harder to park" (185). Haines powerfully articulates the splendor of the physical world in his prose and his poetry, and there are always people inhabiting this splendor and working in and with it, all joined by the poet through memories of those long gone:

> Much rain has fallen. Fog
> drifts in the spruce boughs,
> heavy with alder smoke,
> denser than I remember.
>
> Campbell is gone, in old age
> struck down one early winter;
> and Peg in her slim youth
> long since become a stranger.
> The high, round hill of Buckeye
> stands whitened and cold.
>
> I am not old, not yet, though
> like a wind-turned birch
> spared by the axe,
> I claim this clearing
> in the one country I know.
>
> Remembering, fitting names
> to a rain-soaked map:
> Gold Run, Minton, Tenderfoot,
> McCoy. Here Melvin killed
> his grizzly, there Wilkins
> built his forge. All
> that we knew, and everything
> but for me forgotten. ("Rain Country," *New Poems* 24)

When asked about the intersections of poetry and prose, Haines replied that he likes to think in terms of the German concept of *Dichter*

(Cooperman, "Interview"). Haines describes the *Dichter* as writer, poet, speaker to the people, even prophet. The German term is all of these, and, even more intriguing, it defines one who is a *Schöpfer von Sprachkunstwerken*, a creator of artworks out of language. *Schöpfer*, of course, has a similar meaning to the original meaning of the poet as a maker: one who makes art from words. Haines is a poet and prose artist of ecological perception, is a *Dichter*, one who makes art from words, but also one who makes words from the world, one who sees the world as a place fit for democratic communities of people, landforms, and life-forms, and offers some hope that humans, with some work, may be fit for the world.

Iceberg. Courtesy of the Library of Congress, LC-USZ62-101010.

5
Northern Imagination
Wonder, Politics, and Pragmatist Ecology in Barry Lopez's *Arctic Dreams*

The tundra is a living
body, warm in the grassy
autumn sun; it gives off
the odor of crushed
blueberries and gunsmoke.

In the tangled lakes
of its eyes a mirror of ice
is forming, where
frozen gut-piles shine
with a dull, rosy light.

Coarse, laughing men
with their women;

one by one the tiny campfires
flaring under the wind.

Full of blood, with a sound
like clicking hoofs,
the heavy tundra slowly
rolls over and sinks
in the darkness.
   —John Haines, "The Tundra"

For all art is a process of making the world a different place in which to live, and involves a phase of protest and of compensatory response.
   —John Dewey

On the drive from the Columbus airport to Athens, Ohio, where he was to give a reading, Barry Lopez told me that he kept a copy of John Haines's collection of poems *Winter News* near his bedside. My chapters on Lopez and Haines appear alongside each other not only because both writers have deep ties to the Far North but also because they view that landscape in ways that inform each other. Haines was a resident, Lopez primarily a visitor or journeyer, not, like Haines, focused on a specific place, but providing, as Lawrence Buell writes, "a complementary approach in his major books, *Wolves and Men* and *Arctic Dreams*, which draw heavily on sojourns among northern aboriginal peoples but are not localized anywhere and are inspired as much by intelligent eclectic reading in science, anthropology, and myth as by direct conversation with nature" (*Environmental Imagination* 108). In this sense, Lopez's experience seems to be closer to the kind of experience most of his readers will have of the Arctic. Scott Slovic suggests that "Perhaps it is just as important—and more feasible—for us to use Lopez's reports of the exotic to enrich our understanding of the familiar" (150). That understanding can lead to a "congruence between the self and the world—the pinnacle of intimacy for Lopez—[that] *Arctic Dreams* is working toward" (155). The poem I use as the first epigraph to this chapter comes from *Winter News*, and it suggests how for Haines and for Lopez the Far North is a "living body" both metaphorically and ecologically. Its living ecosystem includes landforms, animals, plants, and people living on the land in often violent and brutal ways—"The land, an animal that contains all other animals, is vigorous and alive" (*Arctic Dreams* 411). The tundra also becomes a trope for how the physical world is part of all living experience. The beauty of survival in and inhabitation of the North, suggested by Haines's infusion of "gut-piles" with a "dull, rosy light," grounds the essays in *The Stars, the Snow, the Fire*. The wonder of the Far North perceived as an intimate collectivity of dynamic processes and the possible consequences of that wonder will be my primary concern in this chapter.

For this reason, I have also chosen an epigraph from John Dewey's *Experience and Nature,* with its insistence on art as a process, because it makes a claim for the work of art as a force for cultural change. Art is useful. It has an ethical dimension. Dewey also claims that there must remain in a work of art a residual "structure of things in environment" (*AE* 101), and these two statements work together—are intertwined. I will continue to investigate ecological writing as an interpenetrative art form, a form accountable to an internal, culturally mediated landscape *and* to an external landscape; therefore, it is a form that is both compensatory (it recuperates the integrity of the external landscape) and an utterance of protest (it denies complete mediation of the natural landscape by cultural forms). Lopez, a writer who consistently demands of himself a responsibility to both cultural and natural structures, insists in *Crossing Open Ground* that "The interior landscape responds to the character and subtlety of an exterior landscape; the shape of the individual mind is affected by land as it is by genes" (65). Or as Dewey argues in *Art as Experience,* "The career and destiny of a living being are bound up with its interchanges with its environment, not externally but in the most intimate way" (19). This interchange is a basic element of experience, and hence of any aesthetic process, either from the artist's or the spectator's position. We cannot lift the mind out of the land, even if the land is a city park or the soil in the rain gutter of my garage supporting a maple shoot.

It follows that ecological nonfiction also protests dominant ways of looking at the world and its human and nonhuman inhabitants through its interrogation of strict distinctions between concepts such as human and animal, animal and mineral, cultural and natural, subject and object. Dewey's notions of protest and compensation are brought into use in Lopez's work through the creation of an ecological aesthetic. Matthew Cooperman writes that a poem "is of itself an ecology" ("Traveling Papers" 25), and ecological writing aspires to the same condition—a pragmatist ecology of world and text. This attempt is most remarkable at permeable boundaries or, as in the work of Steinbeck and Carson, at ecotones. Again, an ecotone is a fluid boundary between ecosystems where gene pools and adaptation strategies are exchanged, whereby evolutionary potential is put into motion. A transformational site of heightened possibility, an ecotone may also include the interface of the material world and products of the human imagination, ecological writing for one. As Romand Coles insists, ecological writing and philosophy can fertilize each other, yielding ecotones of thought: "Most essentially, these concerns call thought to reflect continually upon its own proximity to and distance from the world around it and, furthermore, to locate itself at the interstice between proximity and distance, identity and difference, for it is at those edges that thought is most supple and capable of the care in which its own fertile freedom and that of the world lie. Nothing in the world is more rigid and careless than a thought blind to these edges—these tensions" (237). The primary threat to possible freedoms of thought and world is rigid and careless human thinking that diminishes the art of knowing. The ecological text is involved in the patterning of possibility, and like Peter C. Van Wyck, who sees that "the act of critical interpretation is but the performance of possibility" (2), I concern myself in this chapter with what seems possible for ecological nonfiction in our turn-of-the-millennium world.

The work of ecological writing has—as all art has according to Dewey—a use. Both a work of art and the contemplation of nature similarly exercise an "instrumental function," and "We are carried to a refreshed attitude toward the circumstances and exigencies of ordinary experience. The work, in the sense of working, of an object of art does not cease when the direct act of perception stops" (*AE* 144). The work of art is conducive to an experience that extends beyond the moment of perception. An ecological text can contribute to an ongoing awareness of the place of the human in the natural world, or, again, the ecological text functions "not just in the reader's transaction with it but also in reanimating and redirecting the reader's transactions with nature" (Buell, *Environmental Imagination* 97). The text is, in Dewey's sense, "instrumental."

I want to clarify further Dewey's use of the term "instrumental" here. Dewey writes: "Indeed, persons who draw back at the mention of 'instrumental' in connection with art often glorify art for precisely the enduring serenity, refreshment, or re-education of vision that are induced by it. The real trouble is verbal. Such persons are accustomed to associate the word with instrumentalities for narrow ends—as an umbrella is instrumental to protection from rain or a mowing machine to cutting grain" (*AE* 144). Pragmatist ecology, and Lopez's art in particular, is concerned, certainly, with a "*re*-education of vision," and with "*re*animating" and "*re*-directing" our relations with the natural world. For Dewey, this process prefixed with *re* defines instrumentalism. The process is an ongoing realignment of our experience. Pragmatist ecology and Lopez's art, in this sense, have a "use," are instrumental in the larger Deweyan sense, in which, as Hugh McDonald points out, "Value is primarily connected with activity . . . rather than the object" (108). So value is not necessarily contained primarily within Lopez's work but rather in that work's relation to reader and physical world, in the activity of perceiving relations. With its emphasis on restraint and wonder, Lopez's work enacts a pragmatist ecology. His writing is, indeed, of use to our culture.

To see how this works, I focus on Lopez's National Book Award–winning volume *Arctic Dreams: Imagination and Desire in a Northern Landscape* (1986). I will then glance at political developments in the Canadian Arctic at the turn of the twentieth century and try to imagine a link between an environmental aesthetics and democratic political possibility. If we insist, as I think we must, on the embeddedness of culture in a natural world, we have to extend our critical concerns to that world. There is a world beneath us and our ideas, and it will likely outlive us. Although the terminology may sound at first paradoxical, in order to broaden the scope of our inquiry we may just have to restrain ourselves. Just as it is obvious that we will have to curtail our use of natural resources if we wish our descendants to enjoy life on a green earth, it is becoming ever more clear that the imposition of our Western ways of thinking upon a world that is still largely beyond our knowledge demands restraint. Arguably, the order of the above sentence should be reversed, as our cultural ways of thinking about the natural world allow us to exploit it without restraint. Pragmatist ecological thinking, expression, and action all involve the willingness to limit our use of power and to check our intellectual arrogance, to leave some room for wonder and for the world.

To acknowledge that our own ways of thinking and our own consciousness are necessarily partial is to carve out a space for the world to reveal itself as something shared by many forms of sentience, not only the human. We do not have to know this to be true; we need only know it to be possible. Randall Roorda suggests a "biopragmatist" critical approach that would ask, "What difference might it make in biological terms to take this text, this set of texts, or this mode of commentary as exemplary? What does it mean to seek and accept conversion upon the terms the work provides? What does it entail for the life of the place depicted and that of the places left behind?" (100). Lopez suggests some answers to Roorda's series of questions:

> Walking across the tundra, meeting the stare of a lemming, or coming on the tracks of a wolverine, it would be the frailty of our wisdom that would confound me. The pattern of our exploitation of the Arctic, our increasing utilization of its natural resources, our very desire to "put it to use" is clear. What is it that is missing, or tentative, in us, I would wonder, to make me so uncomfortable walking out here in a region of chirping birds, distant caribou, and redoubtable lemmings? It is restraint.
>
> Because mankind can circumvent evolutionary law, it is incumbent upon him, say evolutionary biologists, to develop another law to abide by if he wishes to survive, to not outstrip his food base. He must learn restraint. He must derive some other, wiser way of behaving toward the land. He must be more attentive to the biological imperatives of the system of sun-driven protoplasm upon which he, too, is still dependent. Not because he must, because he lacks inventiveness, but because herein is the accomplishment of the wisdom that for centuries he has aspired to. Having taken on his own destiny, he must now think with critical intelligence about where to defer. (*Arctic Dreams* 38–39)

Like Dewey, Lopez complicates our unexamined understanding of "use." As in Haines's poem, the tundra is a living process—an Arctic body with the eyes of a lemming, the feet of a wolverine, the voice of birds, and a circulatory system of protoplasm. In the face of the tundra and its creatures, human wisdom is fragile, though human power to exploit remains great. To reiterate Haines, "before knowledge, there was wisdom, grounded in the shadows of a dimly lit age" (*Stars* 157). Instead of using the world up, accepting the terms of restraint, wisdom, and attention in the face of a living world involves deep consequences for how we participate with the world, for the world itself, and for everything else in it.

The operative words for Lopez and for those seeking and accepting conversion to an ecological way of living are "restraint" and "defer," words that in our colonial tradition are usually signs of weakness. As David Orr points out, "in a society that worships technology, questions of this sort are heresy" (39). Lopez claims that we have been and largely remain committed to putting the land and its creatures to use just like

Dewey's umbrella—"you can't have a consumption-based culture unless you have an immoral relationship with nonhuman species" (qtd. in O'Connell 13).[1] This notion of consumption and use is the one established for the Americas from the moment they were colonized, and it is now imperative for the ecological, biological survival of human beings and the rest of the world to embrace the rhetoric of restraint—so unpalatable to our Western ethos—that we see throughout Lopez's work and throughout the ideas of a pragmatist ecology in general. This is how we might put Lopez's work—his quest for intimacy with the land—to use. Accepting his terms of restraint, we can show respect for both otherness and possibility in much the same way that Attridge claims that suspension of habits and known ways of behaving and thinking are key acts in the assumption of ethical responsibility for the other.

Our received ways of thinking about the landscape have become rigidly dogmatic, so much so that they obscure destructive behaviors and allow us at times to perceive destruction as progress. They have become dead metaphors: "And God said, Let us make man in our image, after our likeness: and let them have dominion over the fish of the sea, and over the fowl of the air, and over the cattle, and over all the earth, and over every creeping thing that creepeth upon the earth" (Genesis 1:26).[2] We desperately need different lenses through which to perceive the nonhuman world. The Arctic is especially helpful in this because of its radical difference. As Lopez writes, "this landscape is able to expose in startling ways the complacency of our thoughts about land in general" (*Arctic Dreams* 12). More generally, the extreme otherness of the Arctic forces us to look at ourselves differently—the landscape of the Far North will not accommodate our sedate ways of understanding the processes of the world, and the most obvious example is that our firm conception of the diurnal cycle no longer holds in the Arctic—internalized metaphors of sunset and sunrise no longer make sense in the same way. Certainties are disrupted—the sun does not *always* rise in the east and set in the west. Conceptually more important for this discussion is the fact that it is a place dominated by ice for long periods of time. To appropriate a term from the Canadian geographer Louis-Edmond Hamelin, the *nordicity* of a place measures its degree of difference from a temperate environment, and I am concerned here with regions of extreme *nordicity* (see Bone 18).

Lopez travels the far northern reaches of North America in search of a viable human relationship to landscapes and geographies, even if this relationship often exists primarily in the imagination. According to Thomas Alexander, "As a phase of action, imagination was an essential feature of environmentally engaged rationality or 'intelligence,' as Dewey termed it" ("Moral Imagination" 371). Imagination fuels the art of knowing. Similarly, in an article about ecology and American literature, Karl Kroeber calls for a reinstatement of the imagination in the critical intelligence: "Conceiving culture as participating in an ongoing natural evolutionary dynamic . . . restores the idea that imagination is a fundamental link between ethics and aesthetics. *Imagination* is today a rare critical term because our criticism is controlled by a hegemonic rationality confined within a timeless, structuralist-poststructuralist thinking dependent on explanatory tools such as the model. This essentially Cartesian, mechanistic method cannot accurately represent any existing ecosystem" ("Ecology" 324). The human genome project has strongly suggested that Cartesian thinking fails to account for the complex interrelational structure of our genetic makeup, and Dewey, Muir, Steinbeck, Ricketts, Carson, Haines, and Terry Tempest Williams all reject Cartesian duality and mechanism as insufficient to represent a living world. Lopez does too, and few writers anywhere are more interested in the nexus of aesthetics and ethics than he is. For Lopez, this nexus must be housed in the imagination.[3]

Lopez's method of composition implies an aesthetic and ethical consideration of not just human beings, but all inhabitants of an ecosystem. In an interview with Alice Evans, Lopez describes his compositional strategy in *Arctic Dreams:*

> I chose mammals, like us, across the board, and then with them explored what it means to have a sense of place. . . . I wanted to be able to work on land, then move off the ground onto the ice, with the polar bear going in and out of the water. Then *into* the water with the narwhal. . . . Human beings become part of the landscape through movement, the migration of people into a place. Once I have all these things moving, then I have got fairly well defined what I mean by a "creature," a human being or a muskoxen or me the narrator or you the reader. Animate presence in the landscape. And once that's established, then ice and light, which are in some way ineffable, can be understood as creatures, making and animating the space in which they are moving. The moving icebergs are so beau-

tiful they take your breath away, so beautiful they make you afraid. (Evans 74)

Lopez's imagination provides an aesthetic and ethical link to the nonhuman creatures of the North, and the human becomes a creature on par with a whale or a bear. Not only the animals of the North but "me" and "you" become an "animate presence in the landscape." Our transactions with the natural world are "reanimated." Further, in his imagination, by figuring light and ice as creatures like animals and humans, Lopez is able to extend this consideration to the physical landscape itself, and we all become linked: reader, writer, critic, text, and myriad other creatures and landforms in an ecosystem of the imagination that has aesthetic and ethical import because all members are vitally intertwined.

Although Lopez's use of the imagination in this sense is something new, he is certainly not immune to retooling traditional ways of seeing the landscape. In other words, our inherited ways of seeing will no longer suffice, but this is not to say that they are no longer useful. We see in this passage a clear engagement with the literary sublime through the juxtaposition of fear and beauty. However, Lopez moves beyond the traditional structure of the romantic sublime. True, the incomprehensible beauty of the ice inspires a form of fear of the larger-than or beyond human, but Lopez engages the sublimity abiding in relationships between human and nonhuman consciousnesses, encompassed in what he calls a "creature." The creature emerges from the context of relationships, and to reiterate Dewey's arguments about the centrality of continuity in his aesthetics, "Experience occurs continuously, because the interaction of live creature and environing conditions is involved in the very process of living (*AE* 42). Lopez takes this notion of experience a step further into reciprocity—both the physical world and the human being can be understood as a "live creature." A sense of awe arises from the recognition that our construal of the world is not the only one—you the reader and a muskox exist in a sort of parity in the landscape—and this can indeed elicit from us fear. An ecological sublime.[4] Our easy, and perhaps not wholly avoidable, belief that we are inherently superior to an ox or a bird or an ape is undermined. This destabilization of the self puts in place the nexus of beauty and fear with the added dimension of respect, and a new ecologically sublime relationship to the nonhuman world emerges, one that entails a relinquishment of the self as the rigid center of consciousness.[5] We exist in a Venn diagram of infinitely complex, interlocking, and interpenetrating consciousnesses: to acknowledge this would be to engage a pragmatist ecology that integrates science, moral and ethical questions, the physical world, and art, that resuscitates democratic culture and claims the centrality of environmental values within that culture. A pragmatist ecology takes an outward step beyond our own parameters into the dynamic complexities of ecosystemic experience.

This thought recalls the ecological concept of patch dynamics. Again, patches are units within a larger ecological system that differ in biotic makeup or appearance. A patch could be a burned area in a forest or an entire island. Wu and Loucks argue for a conceptual model of patch dynamics in which "metastability" is often attained. "Metastability" means that because disturbances on the level of patches can be absorbed by the larger system, in many cases instabilities interact in ways that create a relative stability in the larger system: "Nonequilibrium patch processes at one level often translate to a quasi-equilibrium state at a higher level" (453). Part of their conclusion is that "harmony is embedded in the patterns of fluctuation, and ecological persistence is order within disorder" (460). This concept of metastability seems useful in that it can account for all elements within a system as intertwined, each affecting the others, and those interactions *in toto* represent the relative stability of the whole. The paradigm emphatically denies a "balance of nature" view of the world, and in this way allows for conceptualization of the contribution both order and disorder offer to the continuity of ecological systems. Nothing is ever static. The paradigm also revises the notion of the ecotone, enabling us to perceive a system as a mosaic of patches with myriad ecotonal spaces, ranging from the level of genetic structure to the quiltwork of ecosystems that constitutes the Arctic.

Further, all of Lopez's works, as William Rueckert argues, "originate from an awareness of [a] need to reestablish a more harmonious, less destructive, reciprocating, respectful relationship with nature, . . . with the Other—not the alien, not the divine or transcendental, but the wild, . . . the most varied gene pool of all life" (138). Lopez's initial move toward a dialogue with this gene pool represents not only an attempt to envision the world through the eyes of other creatures and other cultures but also an attempt to experience their *Umwelten*, their ecological patches: "The challenge to us, when we address the land, is to join with cosmologists in their ideas of continuous creation, and with physicists with their ideas of spatial and temporal paradox, to see the subtle grace and mutability of the different landscapes. They are the crucibles of mystery, precisely like the smaller ones that they contain—the arctic fox, the dwarf birch, the pi-meson; and the larger ones that contain them, side by side with such seemingly immutable objects as the Horsehead nebula in Orion. These are not solely arenas for human invention" (*Arctic Dreams* 411–12). Ecological patches are nested within each other—species (including humans), landscapes, universe. The human task is to perceive reciprocity and engage in conversation among these ecological levels, to take ethical responsibility for them through attention to interrelationships, especially to the participatory nature of interrelationships. Lopez creates a milieu in which animate creatures and landscapes as well as writers and readers are not only inextricably involved *in* the environment, since this implies a level of separability of the environment and the human, but in which reader, writer, creature, ice, and landscape *are* patches that together constitute an ecology in which human expression is a key participant; in fact, as Lopez will have it, "The very order of the language, the ecology of its sounds and thoughts, derives from the mind's intercourse with the landscape" (278). None of the writers in this study accept the notion that human beings live on the world—we live along with it, are integrated with it—and Lopez's sense of patchy systems seems a promising model for thinking about cultural and natural ecosystems in a global future because it allows for the astonishing diversity of the world without collapsing it into one system.

Lopez's connection with the Arctic is especially important in another sense. *Arctic Dreams* is not only about the ecology of the North and language, but also about global terms marked by the choice of subject matter itself. One of the key points made by Lopez's work on the Arctic is the same one made by Sherill Grace in her essay on Canadian and American literature, "Comparing Mythologies: Ideas of West and North"—namely, that "Rather than being a bounded continental space, the North is open to the world, reminding us forcibly of what the West can ignore—our connection with much of the rest of the world" (250). As Susan Kollin points out, Lopez, by choosing to write about the North, a transnational region, refuses to subscribe to an ecological vision that sees one region, say a wilderness area, as separate from any other region. Because of this sense of circumglobality, Lopez's pragmatist ecology refuses a traditional hierarchizing of natural, scenic wonders (Kollin 67–72). What happens out on the arctic ice—or, for that matter, what happens in Bend, Oregon; Athens, Ohio; or Bonn, Germany—has ramifications important to all ecologies.

Walking along the edge of the ice floe in Admiralty Inlet, Lopez attempts to enter the specific ecology of the place. He meditates on the interstices of ice and open water, and his meditation brings the conceptual edges of the ecosystem into focus. He writes: "This is a special meeting ground, like that of a forest's edge with a clearing; or where the fresh waters of an estuary meet the saline tides of the sea; or at a river's riparian edge. The mingling of animals from different ecosystems charges such border zones with evolutionary potential" (*Arctic Dreams* 123). At these sites of transition, interactions between different systems of creatures and inorganic matter are charged, and any notion of strict demarcation vanishes. Just as the surface of the sea did for Carson and the littoral did for Steinbeck and Ricketts, for Lopez, the ice makes these ecotonal processes visible: "Flying creatures here at Admiralty Inlet walk on ice. They break the pane of water with their dives to feed. Marine mammals break the pane of water coming the other way to breathe" (123). Not only the ice, but also the ecology of Lopez's language aid the reader's perception of the ecotone. Even though borders between ecosystems are abstract, fluid, difficult to comprehend, Lopez's repetition of the phrase "pane of water" concretizes this very abstract notion for the reader. Ice provides the membrane between water and air, and the image dramatizes the action of the creatures penetrating these borders. Breaking the pane of water, like smashing a window, is a dramatic metaphor that suggests, perhaps, the shattering of preconceived notions, of received ideas about our complete difference from the rest of the world. The North, with its preponderance of ice, provides Lopez with a place where he can actually see, touch, and report upon the idea that drives his ecological aesthetic, namely, that beauty is inherent in the ecotones where the future of life on the planet is in the process of generation. In this sense, according to Sherman Paul, "Aesthetics, which hitherto involved the visual appreciation of nature, . . . the spectatorial viewing of its scenery, is now a cover-word for depth of ecological understanding" (*Essays* 98). In this view, aesthetics inheres in ecological perception and understanding. Lopez's location in this particular landscape places the writer and the reader within a nearly visible web—on the perceptual fringes—of evolutionary potential, gathering together the strands of an ecological aesthetic.

This gathering happens at the edges, or as we have seen in Stephen

Tatum's construction, at "topographies of transition" whose borders are "membranes" through which seemingly oppositional elements interact (325). Tatum, again, levels the ontological, epistemological, and ecological and erases the privilege of any one category. Ontology and epistemology by definition refer back to the human; ecology insists upon an integration of the human with the nonhuman, and in so doing it breaks down the borders drawn between human life and its environment. Grace writes that "border lands border on borders, exist profoundly at edges, thereby blurring, blending into, and spilling over on all sides" (243). The term "edge" is not only a general term but one used with a specific meaning in the language of ecology. An "edge effect" is the pulse, the pushing and pulling and spilling over that happens at the edges where ecosystems meet, "a tidal ebb and flow of energies that, in its middle passage as well as its extremes, nurtures characteristic ecosystems" (Elder, *Imagining* 192). In this sense all boundaries are ecotonal—edges, places of enhanced evolutionary potential for live creatures, habitats, texts, sensations, knowledge systems, and beliefs—and the spilling over at these boundaries suggests the continuity, the ongoing nature of experience. Lopez continues along these same lines: "That attraction to borders, to the earth's twilit places, is a part of the shape of human curiosity. And the edges that cause excitements are like these where I now walk, sensing the birds toying with gravity; or like those in quantum mechanics, where what is critical straddles a border between being a wave and being a particle, between being what it is and becoming something else, occupying an edge of time that defeats our geometries" (123). Ecotones not only breed evolutionary potential but also power the art of knowing, primarily through stimulating the human imagination to perceive relations in process. These places, again made visible by the ice and by Lopez's prose, help to "shape" the imagination, and the choice of the verb "shape" is a further attempt to bring the abstract into focus, as is the reference to geometry, the attempt to measure the relationships, say, between lines, points, surfaces, and angles, the attempt to graphically represent abstract points of contact like planes and edges. Lopez also includes the language of quantum physics, whose liminal particles have so captured our imagination, but just as important here is the link between quantum physics as a marker of the historical moment when scientific discourse begins to lose its stability and the therefore tacit recognition that Lopez's discourse on the Arctic, symbolized by the very ice upon which he stands, cannot be other than a fluid stance, one that must change and evolve as it interacts with landscapes and peoples far removed from our own temperate zone imaginations (see Papin 1254). Reflecting on Heisenberg and the uncertainty principle, Dewey writes: "The principle of indeterminacy thus presents itself as the final step in the dislodgement of the old spectator theory of knowledge. It marks the acknowledgement, within scientific procedure itself, of the fact that knowing is one kind of interaction which goes on within the world" (*QC* 163). Knowing is not directed at the world from without; it is part of what happens in the world, and it participates in the world. As early as 1929, Dewey refuses to rule out the possibility that there are other ways of interacting with the world aside from human knowing. Knowing is never finished but always ongoing, never solitary, always an interaction with an environment.

Lopez's invocation of the uncertainty inherent in particle physics reveals a concern that runs through *Arctic Dreams*. A renewed focus on the shifting, patch-like relationships among the nonhuman world, contemporary human imagination, and inherited cultural structures, including scientific inquiry, is central to all of Lopez's work. Along with T. F. H. Allen, Lopez understands that science can produce not only equations but also powerful metaphors: "Of the sciences today, quantum physics alone seems to have found its way back to an equitable relationship with metaphors, those fundamental tools of the imagination" (*Arctic Dreams* 250). There no longer exists a dichotomy between science and the fundamental tools of written art, metaphors. In his study of Dewey's philosophy of value, James Gouinlock places the erasure of this duality at the center of Dewey's thought: "Of fundamental import in Dewey's theory of nature is his insistence that the world as disclosed by experience and the world as disclosed by science no longer present an unbridgeable dualism, but are continuous with one another" (67). Lopez extends this continuity by linking quantum physics to the cathedral culture of the High Middle Ages, then taps the medieval interpretation of the Greek concept *agape*, and finally teases out its definition to signify "a humble, impassioned embrace of something outside the self, in the name of that which we refer to as *God*, but which also includes the self and *is* God" (*Arctic Dreams* 250). Here, Lopez's admission of uncertainty is marked by the term "something," and the syntax of his metaphor reimagines a traditional monotheistic *God* as the "impassioned embrace of something outside the self" that also includes the self. The impassioned *relationship* between self and world replaces a traditional God. No longer nominal, God is verbal, in motion, in experience. Throughout the narrative, Lopez reworks traditional structures in order to help ground his ecological aesthetics.

In another context, Lopez claims along the same lines, "I think if you can really see the land, if you can lose your sense of wishing it to be what you want it to be, if you can strip yourself of the desire to order and to name and see the land entirely for itself, you see in the relationship of all its elements the face of God" (qtd. in Aton 16). Again, "God" for Lopez is a matter of relationships among all the elements of the land. A recognition of those relationships requires the stripping away or refocusing of human desire from colonial order toward maintenance of interrelations, a position that modifies its obvious Christian overtones in order to embrace a more ecological focus.[6] His work is not a throwing away, but a reimagining—a pragmatist ecology—of our shifting relationships to spiritual matters, to science, to land, to ice, and to culture.

Ice always shifts. It melts at its edges, shifts, buckles, and drifts, so it can provide only a boundary in flux. The area around Lancaster Sound teems with wildlife, staggering in its numbers and reminiscent of pre-contact North America; however, "It is the ice . . . that holds this life together. For ice-associated seals, vulnerable on a beach, it is a place offshore to rest, directly over their feeding grounds. It provides algae with a surface to grow on. It shelters arctic cod from hunting seabirds and herds of narwhals, and it shelters the narwhal from the predatory orca. It is the bear's highway over the sea. And it gives me a place to stand on the ocean and wonder" (*Arctic Dreams* 124). The ice floe provides the human being with a platform from which to see, speak, and participate. The writer and reader arrive in this complex world of seals, birds, ice, algae, whales, and polar bears, and in this scene the artist performs a specific function in the ecosystem. Indeed, for Lopez, the ecosystemic role of humans may at times simply be our capacity to stand and wonder, to wonder ourselves into awareness that we are part of an intricate, grand system of existence, not simply creatures who stand in separate awe of it. Of course, our involvement in ecosystems has often been apocalyptic for other species, and ecological writing works both "To cultivate in your own mind the awe of an adult, to be aware of how difficult and black human life can be. But at the same time, to cultivate a sense of wonder" (qtd. in Bonetti 66). Along similar lines, Dewey, early on in *Experience and Nature,* contends that "We cannot achieve recovery of primitive naïveté. But there is attainable a cultivated naïveté of eye, ear and thought, one that can be acquired only through the discipline of severe thought" (40). What Dewey means here is that he intends to refocus philosophy onto "the things of ordinary experience." That is what his and Lopez's rigorous naïveté is about. Their recuperation of the wonder of ordinary experience resonates with many of us concerned with environmental degradation and just as alarmed at the threatened state of democracy and the influence of unprincipled wealth as Dewey was nearly eighty years ago. This absence of principled life threatens the highest aim of Dewey's thought as formulated by Alexander: "the democratic community for Dewey is the community which understands itself as creatively pursuing life as art" ("Pragmatic Imagination" 341). More than anything else, this lack also threatens the aims of pragmatist ecology. As I have argued at length, art depends upon the physical world and the community. Lopez's application of severe thought to the interrelationships among humans, land, and animals yields wonder at the world and his own participation in it, suggests the approach of life to the condition of art.

One might argue that such a focus on land and animals is simplistic, or, in Joyce Carol Oates's opinion, "It inspires a painfully limited set of responses in 'nature writers'—REVERENCE, AWE, PITY, MYSTICAL ONENESS" (236). These responses, on the contrary, seem in no way limited. A reworking of open-mouthed rapture when faced with a mountain range, for example, is tedious enough. However, wonder can indeed be a constructive response, especially when it results from an ecological perception of interrelationship. Tony Tanner traces the use of wonder in American literature and comments that American writers "wished for words which would register the circumpressure of the experienced world, words full of the shape of things, heavy with the weight of things," reflecting their focus on the physical world (13). In turn, "wonder and sympathy effectively produce empathy," which implies a heightened level of care for people and things (76). Wonder at one's inextricability from human and nonhuman others in both an ecological and a social sense, the realization that in the degradation of others we diminish ourselves, seems to me in no way "painfully limited." Lopez's work insists upon an open dialogue with the physical world: "To have no elevated conversation with the land, no sense of reciprocity with it, to rein it in or to disparage conditions not to our liking, shows a certain lack of courage, too strong a preference for human devising" (*Arctic Dreams* 412). And too narrow a focus on human constructions can obscure the complexity of human culture's embeddedness in patterns of ecosystems. With Oates, I am quite convinced that we can live without mystical oneness, but we can certainly still benefit from wonder, reverence, awe, pity, and an acknowledgment of the importance of these emotional responses in relation to the natural and social world. Of course, beneath her skepticism one senses a modicum of awe even in Oates at her perception of the nature of her own body: "a tall column of light and heat" (243). Just as that "tall column" insists on its own rigorous and useful thought, Lopez stands at an ecotone and wonders at the world, and instead of seeing wonder as a form of banality or as a category, he reimagines it as rigorous and useful—transformed into active participation in the push and shove of ecosystemic potentiality.

This way of thinking about wonder pushes it to the borders of the ethical. Lopez is known as a man who bows to birds and who stops his car to move roadkill from the highways he travels. Whatever we may think of this practice, it is for him a nod of respect and empathy toward animal life—life, as Romand Coles argues, we fail to grant ethical consideration, with the consequence that "as we obliterate the wild nonidentical textures of our world, we simultaneously reduce the potential richness of our own being—beings that are entwined with this world" (231). Wonder contributes to the art of knowing and to the construction of a pragmatist ecology. Humility, awe felt upon awareness of other forms of sentience, and the desire to embrace those forms while not fully relinquishing the self work together to elicit wonder at a world filled with the potentiality of otherness. The world's otherness contributes to the potential richness of our own being that Coles mentions. We feel wonder in the face of both self and world, or at world and self face to face, but only if that wonder is rigorously construed as a form of respect, rather than, say, a pat response to a polar bear killing a seal on the television screen, only if wonder entails a caring for and a heightened awareness of the value of nonhuman life and landscapes.

Lopez worked for a time as a wildlife and landscape photographer, and he recalled an incident in which a polar bear emerged from a snow squall and during which Lopez felt he squandered part of the experience by fiddling with focus and f-stops. The encounter led to the end of his career as a photographer and to the following realization: "I view any encounter with a wild animal in its own territory as a gift, an opportunity to sense the real animal, not the zoo creature, the TV creature, the advertising creature. But his gift had been more overwhelming. In some way the bear had grabbed me by the shirtfront and said, Think about this. Think about what those cameras in your hands are doing" ("Learning to See" 76). Uneasy with placing a frame around the animal's experience, Lopez felt that he could no longer face the world of the bear or the world of the seal through a viewfinder only. Through the imagination—the act of writing activated by the act of reading—the valence of wonder is also turned back toward the subject, and the wonder evoked by the nonhuman world and our own capacity for wonder become intricately related.[7] When we come to see that degradation of the one is inherent in the degradation of the other, we grow more inclined to expand our ethical considerations to the land itself. "The land ethic," as Aldo Leopold urged in 1949, "simply enlarges the boundaries of the community to include soils, waters, plants, and animals, or collectively: the land" (204). As Jen Hill points out, "The definition of community as an expression of a moral and ethical framework is apparent from the very beginning of *Arctic Dreams*" (135). An extension of the concept of community to encompass not only the human but also the land would represent voluntary restraint imposed on received notions of human dominance, and the potential embrace of a living world full of wonder.

At another ecotone, this time a built one—a drilling platform, not an ice floe—Lopez is surprised by a seal surfacing in the "moon pool," the open water beneath the platform: "What held me was: how far out on the edge of the world I am. A movement of my head shifted the hood of my parka slightly, and the seal was gone in an explosion of water. Its eyes had been enormous. I walked to the edge of the moon pool and stared into the dark ocean. I could not have been more surprised by the seal's appearance if it had fallen out of the winter sky overhead, into the spheres of light that embraced the drill rig and our isolated camp" (*Arctic Dreams* 13). Lopez's wonder is here expressed by a heightened attention to detail, the seal's eyes, and by the intersection of detailed description with literary language such as "explosion of water" and "spheres of light." The seal's enormous eyes express its world in the face of massive human-made machinery imposed upon its habitat. What Lopez accomplishes here is the transmission through literary expression of the wonder of the seal's world, and the wonder at his own world made more compelling by the world of the seal. Even the oil rig is bathed in light, perhaps at the moment of his perception, but certainly in the moment of reading. So Lopez's point here is not a call for a return to a more pure past but rather a call to imagine, in the present, future possibilities that are inclusive and respectful of the world of the seal and the human world. (This is not to dismiss the values of the past; as we have seen, Lopez, like Haines, is just as willing to use the past as a critical tool for evaluating the present, his use of *agape*, for instance.) For Lopez, these worlds—these patches—can no longer be separated; they are ecosystemic, and the desired consequence of this formulation is the enhanced awareness of the importance of human action and expression in ecological relationships tending to metastability. We, of all participants, can enact restraint, and restraint may just be the key to our future participation in the world.

Arctic Dreams is deeply concerned with consequences, and that attention to consequences demands heightened awareness of our actions and our situation in the world. Out on the rig, Lopez realizes that "To contemplate what people are doing out here and ignore the universe of the seal, to consider human quest and plight and not know the land, I thought, to not listen to it seemed fatal. Not perhaps for tomorrow, or next year, but fatal if you looked down the long road of our determined evolution and wondered at the considerations that had got us this far" (13). To ignore the consequences of current actions and current structures of thought will be fatal. To quote Graeme Wynn, we need to re-envision our "use of a limited utilitarian calculus" (12). *Arctic Dreams* has, of course, a use; it brings us into an ecotonal relationship with the North's fluid boundaries—it encourages an awareness of our own place in the world. According to Scott Slovic, this "idea of comprehensive awareness as a prerequisite for enlightened behavior seems appropriate as a guideline not only for human involvement in the Arctic, but for our presence in the natural world generally" (141). Part of our presence, if we pay attention to the consequences of our actions, is the realization that not only our physical presence but also the expression of our imaginary presence through language is essential to a viable future relationship with the nonhuman world. Science, art, and world all participate in the art of knowing.

We can imagine ourselves, through language, in a social, ethical relationship with nonhuman existence. Jen Hill identifies how Lopez's prose enacts this relationship: "In *Arctic Dreams* he demonstrates the deep connection between the human and nonhuman, not only by investigating land-centered native practices and knowledge but also by employing a vivid, moving first-person narrative that invites his readers to participate in this deep connection" (134). Lopez writes: "When we enter the landscape to learn something, we are obligated, I think, to pay attention rather than constantly to pose questions. To approach the land as we would a person, by opening an intelligent conversation" (*Rediscovery* 36).

To engage the land as we would other people implies extension of ethical consideration to the land. Lopez, of course, assumes much by suggesting that we habitually approach other people with respect and intelligence, but this too may be exactly his point. A lack of respect and restraint toward other people and toward the land may be a general symptom of our culture. He suggests not only that we approach the land as a person but that we approach it as an intelligent person, which means, of course, that we expect an intelligent answer. This notion is linked to Slovic's "awareness": if we accept that the world has some sort of intelligence, even metaphorically, we remain aware of and alive to the possibility of learning something. Certainly the physical world functions in ways we will never understand, so to deny the possibility of an intelligence because, frankly, we have no way to describe it or to talk about it is to shut off an entire realm of possibility. Such a denial seems an abysmal lack of imagination on our part, a failure of the creative intelligence. If the land has intelligence to offer, it requires a respectful attitude toward it. Lopez suggests that we enter into such a relationship with the land. This relationship entails human acceptance of full responsibility for the predictable outcomes of our behaviors and actions, and full acceptance demands awareness, respect, and restraint.

This comprehensive awareness inheres in a pragmatist ecology, and it is also requisite for Euro-American involvement with other cultures. We have tended to view cultures existing close to the land—and the intelligence they have to offer—with the same disregard with which we continue to view the land, except perhaps in overly romanticized perceptions of noble savages and landscapes. Lopez counters the unexamined romanticization of native cultures. About his extensive interactions with native peoples, he comments: "They are less and less interested in you the more and more interested you are in being like them—because they know you can never be like them. What they wish is that you would express, with the integrity of your own positions in a discussion or in the way you live, the best of what your culture represents. Then there is something to talk about" (qtd. in Lueders 15). Again, Lopez insists upon the retention of valuable concepts made available by one's own culture. The attempt to "go native" is, finally, self-indulgent. Lopez comments, "Today we talk about American Indians or one culture or another being the one we should all aspire to imitate. This isn't helpful" (qtd. in Aton 10). We can certainly learn but not imitate, and perhaps the best way to learn is to think of ourselves as participants, not observers.

Like native cultures, "the land," as Lopez sees it, "retains an identity of its own, still deeper and more subtle than we can know. Our obligation toward it then becomes deceptively simple: to approach it with an uncalculating mind, with an attitude of regard. To try to sense the range and variety of its expression—its weather and colors and animals" (*Arctic Dreams* 228). Notions of obligation and regard have been reserved for human beings, but Lopez insists on extending ethical consideration to the land itself. Dewey too insists upon the reciprocity with the physical world entailed in ideas of obligation and regard. Of course, he did not live to see what humans had done to the physical world by the opening of the twenty-first century. Many native peoples of the North do possess "a timeless wisdom" that insists upon just such an ethical regard for the land. Furthermore, Lopez approaches their wisdom with the same quality of restraint that he exercises toward the land: "It is . . . a wisdom not owned by anyone, not about which one culture is more insightful or articulate" (298), and here he encourages careful cultural exchange. Our regard for other lives is expressed through restraint and tentative approach. Lopez finds both human cultures and the world more often than not beautiful, and he defines beauty as the full exercise of life (89). Beauty, for Lopez, is then expressed through the life of the world. Infringement upon the rights of people, cultures, and things fully to exercise life curtails the processes that bring beauty into being.

Cultural exchange must be dialogic, and intercultural dialogue can be imagined as ecotonal—an interface not only of species and nonliving world, but with the added dimension of human cultural interaction. Here again we can bring Dewey to bear on the relationship between culture and world: he claims in *Art as Experience* that "culture is the product not of efforts of men put forth in a void or just upon themselves, but of prolonged and cumulative interaction with environment" (34–35). Culture participates with its environment, and if we can insist upon an ecological aesthetic, then, perhaps, an attempt can also be made to apply notions of ecology to both cultural exchange and political negotiations. While we too often think about environmentalism as an opposition between those who would preserve ecosystems and those who would destroy them, Susan Kollin calls attention to another level of environmentalism. She makes an astute observation on indigenous environmentalism: "indigenous environmentalism in Alaska understands that conquest and genocide are aspects of a postcontact ecosystem. Indigenous environmentalism also resists understanding the subject and agent of nature writing as a solitary individual in retreat and instead concerns itself with the collective community" (71). As we in the United States inhabit a tradition that has valorized the individual in the wilderness over and against community relationships with the natural world, it seems time that we engage in a dialogue with the many other ways of perceiving the world that insist upon ecologies that include human and nonhuman communities and landscapes. Remember, as early as 1929, Dewey refused to grant our way of knowing exclusive access to the world (*QC* 165). To this end, we need to recognize and care about the effects of our actions not only on ecosystems but also on the communities that are the potential sources of renewed democracy. Lopez writes: "The real topic of nature writing, I think, is not nature but the evolving structure of communities from which nature has been removed, often as a consequence of modern economic development. It is writing concerned, further, with the biological and spiritual fate of those communities. It also assumes that the fate of humanity and nature are inseparable. Nature writing in the United States merges here, I believe, with other sorts of post-colonial writing." The nature essay "is in search of a modern human identity that lies beyond nationalism and material wealth" ("Shaped" 1). Lopez's work always seeks both to recuperate values that have been lost—the removal of nature from community consideration—and to imagine how restored communities might look in cultures that have moved beyond nationalism and materialism and how an ecological identity might be grounded. His work, and works by Dewey, Muir, Steinbeck, Ricketts, Carson, Haines, and Terry Tempest Williams, is "the emergence of a concern for the world outside the self," the awareness that "Nature is not scenery, not a warehouse of natural resources, not real estate, not a possession, but a continuation of community" ("Shaped" 1, 11). Allen defines community as "the interaction, not the physical presence of organisms or populations in a place" (323), and, like Lopez's God and Dewey's experience, community becomes a verb, an active thing in that it is predicated on relationships. If, following Allen's suggestion, we take a metaphorical jump and use this definition to think about human communities, of foremost concern for any individual would be awareness and maintenance of interactions and interrelationships, which, again, would demand an ethical consideration of human and nonhuman others as participants in experience.

It should by now be clear that strict demarcations between human and nonhuman, self and world, pre- and postcontact cultures are, finally, too easy. These distinctions do not help us to perceive the world in a democratic, ecologically sound way, and they can lead to an atrophy of the power to imagine ourselves as participants in a rich and wonderful reality—as well as, of course, in a world equally dark and frightful. Our assertion of power and dominance is a weakness, a deprivation of others' rights and abilities to freely evolve, a denial of potential human experience at a juncture with otherness, in Coles's words, with the "nonidentical textures of our world"; it is a foreshortening and boxing in of our self-potentiality. It stifles democracy. In *Art as Experience*, Dewey complains about mental compartmentalization that would allow the separation of theory from practice, of emotion and imagination from action:

> Since religion, morals, politics, business has each its own compartment, within which it is fitting each should remain, art, too, must have its peculiar and private realm. Compartmentalization of occupations and interests brings about separation of that mode of activity commonly called "practice" from insight, of imagination from executive doing, of significant purpose from work, of emotion from thought and doing. Each of these has, too, its own place in which it must abide. Those who write the anatomy of experience then suppose that these divisions inhere in the very constitution of human nature. (26–27)

Of course, they do not inhere in human nature. This compartmentalization degrades the art of knowing at all levels of intercommunication, and indeed, for Dewey it represents a withdrawal from the world. Dewey, of course, sees these activities as interrelated. Practice cannot be divorced from insight (or theory) any more than thought can be removed from emotion, or emotion from doing any more than culture can be lifted from the land. Human cultural endeavor (in all its forms) and the natural world exist in a pragmatist ecology, and for both Dewey and Lopez, once these embedded conceptual boundaries are experienced as loosened and fluid—in other words, experienced as ecotonal—experienced as places of heightened evolutionary potential for living things and for cultures, then one arrives at "a transformation of interaction into participation and communication" (*AE* 28), which also echoes Poirier's transformational moment of wonder and beauty (202). It also puts democracy into motion. So, the attention to ecotones—or the willingness toward interrelation—is the prologue to meaningful dialogue and participation in all forms of endeavor: in art, in work, in academic disciplines, and in democratic politics.

༄

On a hopeful though still problematic note regarding the political interface of cultures and democracy at work, on 1 April 1999 the Canadian territory of Nunavut came into existence, and this political advance, although it is by no means a perfect deal, gives the Inuit of the Canadian Arctic virtual political control of their own territory. Nunavut, which translates as "our land," constitutes a landmass larger than any other Canadian province or territory and represents perhaps the largest and richest native land claim in the world (Pelly 22). Eighty-five percent of its twenty-two thousand residents are Inuit, and the official language of government is Inuktitut. Almost half of the population is under the age of twenty, a fact that is two-edged: it could yield great potential, but will surely result in lack of employment opportunity and a related out-migration. The terms of the Nunavut Final Agreement, which calls for an eventual 85 percent Inuit participation in civil service posts and the creation of "birthright corporations" (corporations collectively owned by the Inuit population of Nunavut), represent meaningful steps toward an economy infused with "traditional Inuit values and skills, as well as the talents and energies of relative newcomers" (K. Harper). Elders are included in native-language instruction and the preservation of native culture, and the legislative assembly will attempt to maintain a consensus-based decision-making process modeled on aboriginal ones (Vlessides).

In the context of this study, what interests me here so greatly is that, as Peter Usher writes, "The single most important factor in the course of northern development in the last twenty-five years has been the assertion of power by aboriginal northerners, and the accommodation of Canada and its two northern territories to that assertion. Underlying this reversal was the simple but underestimated . . . survival of aboriginal distinctiveness; of economic, social, and cultural persistence of a way of life and of a commitment to a different vision of the future" (379). The Inuit have insisted on their *nordicity*. Their insistence upon the integrity of a way of life as it is pushed against by a more powerful way has resulted in an exchange that can be preliminarily seen as positive, and as Elke Nowak writes, one of the highest cultural accomplishments of the Inuit is their ability to adapt to one of the harshest environments on earth, an accomplishment reflected in their determination to control what aspects of the dominant culture they decide to put to use for the ongoing development of their own culture (29). Certainly, the economy of Nunavut will have to take into account global capitalism, but Brian Calliou and Cora Voyageur argue that "such understanding does not mean accepting and reproducing the capitalist way of doing business. To avoid the political, social, and cultural assimilation that would follow such economic assimilation, Aboriginal organizations must be selective, adapting and modifying existing capitalist business practices to meet their cultural beliefs and needs" (131). In this sense it is unwise to look at their culture as static, as being penetrated by the more powerful one to the south.

At the interface of two cultures there is exchange, and like the ecotone where ecosystemic exchange happens, there is evolutionary potential. As Usher suggests and Nowak writes, the insistence of the Inuit on the maintenance of their culture while filtering in aspects of the dominant one is "not only a repudiation of a forced isolation, but it can be seen as a lesson for the Canadian concept of multiculturalism" (20, my translation). Inuit insistence upon a different vision of the future may be a lesson for all of us. If a pragmatist ecology has an ethical valence capable of recuperating the landscape, then it should be clear that similar considerations need to be applied to different cultures. In their over-twenty-year effort toward self-government, the Inuit have insisted upon just such respect.

Now, I do not mean to assert a direct link between American ecological writing and the efforts of the Inuit. That would do justice to neither. But the renewed attention to ecotonal relationships between the human and natural world, and between human cultures, all necessarily embedded in the nonhuman environment, can evolve into participation and dialogue that also insist upon an ecological element of restraint and respect. At the end of *Arctic Dreams,* Lopez expresses his own deep regard: "I looked out over the Bering Sea and brought my hands folded to the breast of my parka and bowed from the waist deeply toward the north, that great strait filled with life, the ice and the water. I held the bow to the pale sulphur sky at the northern rim of the earth. I held the bow until my back ached, and my mind was emptied of its categories and designs, its plans and speculations. I bowed before the simple evidence of the moment in my life in a tangible place on the earth that was beautiful" (414). The integrity of this place and the integrity of cultures who have beheld it respectfully hold out to us a wealth of possibility. In line with Lopez's thinking, but with an added note of dread and warning,

Margaret Atwood insists that the northern imagery so important to the Canadian imagination is "erected on a reality; if that reality ceases to exist, the imagery, too, will cease to have any resonance or meaning except as a sort of indecipherable hieroglyphic. The North will be neither female nor male, neither fearful nor health-giving, because it will be dead. The earth, like trees, dies from the top down. The things that are killing the North will kill, if left unchecked, everything else" (116). We do well to heed Atwood's warning. Part of what is killing the North is, of course, our inability to restrain ourselves. Viewed through the lens of a pragmatist ecology, political reality in the Far North and an ecological creative intelligence alert to ecotonal possibility may yield a healthier way of perceiving ourselves and the world—may yield evolutionary potential—may yield Dewey's "process of making the world a different place in which to live." Ecological writing seen through this lens moves to restore to us the power and beauty of the natural world in all its complexity, and at the same time resists complete absorption of the world into our own calculus of exploitation. The continued integrity of such a world and the people in it make Lopez's bow, and our wonder, a possibility.

Wood ibis, scarlet ibis, flamingo, white ibis, by Alexander Wilson. Courtesy of the Library of Congress, LC-USZ62-52588.

Conclusion
(Eco)logic in the Utah Landscape

If literacy is driven by the search for knowledge, ecological literacy is driven by the sense of wonder, the sheer delight in being alive in a beautiful, mysterious, bountiful world. The darkness and disorder that we have brought to that world give ecological literacy an urgency it lacked a century ago.

—David Orr

Common things, a flower, a gleam of moonlight, the song of a bird, not things rare and remote, are means with which the deeper levels of life are touched so that they spring up as desire and thought. This process is art.

—John Dewey

The examination of environmental problems and the role pragmatist ecology can play in an intervention into dominant ways of thinking about and perceiving the physical world results in a balancing act between wonder and despair. That despair is converted to wonder and hope only through the imagining and the practice of solutions. I am convinced that the most basic step in this process is to alter how humans perceive our relations with the physical world. Until we understand ourselves and our cultures as part of the ecological processes that sustain all life on earth, despair and grief will likely retain the upper hand. Certainly, as I have been arguing, a major part of any solution lies in our art forms and in rendering porous the barriers that impede the art of knowing. Renewed attention to common experience—to a lily on a mountain, the life in a tide pool, the curve of a bird's wing, the glint of mica in a trowelful of mud, the light of the northern sky, a gleam of moonlight, or a family in the Utah landscape—holds the key to ecological survival. All of the writers examined in this study understand that experience ecologically defined must include the common things of our existence. All of their works are emplaced. Still, in our culture, the common things most often taken for granted seem to be the very real places in which we live. Where I live one can hardly travel a mile without encountering a new, nearly identical subdivision, very often a gated one. The search for a renewal of community and place becomes difficult indeed.

A colleague and I were looking over the program for an upcoming conference, and we paused over something that many in academic circles take for granted. We remarked in particular upon the title of the first session: *"Real" and Constructed Space*. The scare quotes around "real" imply, of course, that constructed space is more "real" than "real" space, maybe even imply a knowing wink toward those, presumably the participants in the conference, who are in on the knowledge that nothing outside human discourse is real. I will look to Terry Tempest Williams for help in my closing dissent from this view. Williams inquires into our understanding of the relationship between human beings and their physical environments and emphasizes the realness and particularity of place. In this she is close to John Haines in that she relies upon a long-inhabited place as ground for her spatial and ecological perception. She is unlike Haines, who only briefly mentions the two women who shared much of his experience in Alaska, in that she includes her entire extended family in her ecology. Family and physical place—the social and the natural—are inextricably, ecologically linked.

This heightened attention to place does not simply jettison the constructed nature of it, but insists that things are not always reducible to social constructs although social construction helps to form ideas about things. In *Writing for an Endangered World,* Lawrence Buell calls for a "mutual constructionist understanding of placeness" (16).[1] By this he means that because nature textualized is a cultural construction is not grounds to dismiss the possibility that "attachment to place" can be "a creative force" (17). Further, he argues in *The Future of Environmental Criticism* that "The concept of place also gestures in at least three directions at once—toward environmental materiality, toward social perception or construction, and toward individual affect or bond" (63). Williams bows in all three directions. Through her insistence on the reality of natural places—the Utah landscape in particular—she foregrounds the idea that ecological perception as a key component of meaningful experience is rooted in material places. That rootedness can create "respect for environment as destabilizing force" able to intervene and revise socially constructed ideas (Buell, *Endangered* 17). As we will see, Williams's place is itself unstable, and her giving voice to that place in turn destabilizes our complacent attitudes toward the ecologies that keep us alive both physically and imaginatively. Along these lines, Dewey argues that "Because every experience is constituted by interaction between 'subject' and 'object,' between a self and its world, it is not itself either merely physical nor merely mental, no matter how much one factor or the other predominates" (*AE* 251). His use of "merely" is of great interest here: no experience is complete without the interaction of thought and material environment. One without the other is mere, is diminished. Full creative thinking needs a place, and a place needs fullness of thought.

This idea is furthered by anthropologist/philosopher Edward S. Casey, who convincingly argues that the concepts of "space and time are themselves coordinated and co-specified in the common matrix provided by place. We realize the essential posteriority of space and time whenever we catch ourselves apprehending spatial relations or temporal occurrences in a particular place" (37). We are all of us—even literary critics—"emplaced," and "The world comes bedecked in places; it is a place-world to begin with" (43). In this closing section I will move out from Casey's contention and, drawing on Dewey's logic, examine how Williams contributes to a discussion of the relationships among space, place, ecology, and culture. To my mind, her work provides a jumping-off point for twenty-first-century pragmatist ecology. Williams's importance to this project—and to ecological thinking in general—is that she provides a logic of interconnections. She provides a model for thinking about the human role in natural communities that foregrounds relations of all kinds. These relations are vitally important to her problem-solving strategies, to her (eco)logic.[2] Rejuvenating democracy and solving environmental problems will require strategies of interrelationship among people, places, academic disciplines, and cultural forms. This effort to honor interrelationships, and the communities they create, is central to all of Williams's thought and work as a writer and an activist. It helps that, like Rachel Carson, she has a strong background in the disciplines of English and biology.

Williams organizes her 1991 book *Refuge: An Unnatural History of Family and Place* around the rising lake levels of Great Salt Lake as it inundates the Bear River Migratory Bird Refuge. Place becomes even more specific as the lake is aligned with the cancer slowly killing Williams's mother. In *Refuge,* a particular place is aligned with a particular body, a connection that becomes especially interesting when Williams engages an (eco)logic that can contribute significantly to a dialogue that attempts to articulate ways in which human beings might interact with the natural world without killing it and us off by the beginning of the next century. *Refuge* begins to articulate a pragmatist ecology for the future. It encourages a reformulation of how an aesthetic representation of experience in which natural places play a key role can lead to our perceiving places, spaces, and living communities in a more ecological way.

Simply asked, then, might how we read and write have consequences for "real" places?[3] A concern with "experience" and "consequences," of course, points back to Dewey. Dewey's logic can be brought to bear on the articulation of a pragmatist ecology generally and, more specifically, on how *Refuge* can contribute to that articulation. Dewey insists that experience "occurs continuously, because the interaction of live creature and environing conditions is involved in the very process of living" (*AE* 42). This basic notion of experience as a living process that involves both self and environment as participants is centrally important to what Dewey, in *Logic: The Theory of Inquiry,* calls the "existential matrix of inquiry." Sidney Hook writes that "Dewey's approach takes its point of departure from the fact that thinking and therefore logical thinking makes a difference to the world." Hook then goes on to say that Dewey's logic "recognizes the continuity between man and nature" (xiv). Obviously, inquiry cannot happen without the rigorous application of intellect, but the existential matrix in which inquiry is performed is more extensive than the strictly human:

> Whatever else organic life is or is not, it is a process of activity that involves an environment. It is a transaction extending beyond the spatial limits of the organism. An organism does not live *in* an environment; it lives by means of an environment. Breathing, the ingestion of food, the ejection of waste products, are cases of *direct* integration; the circulation of the blood and the energizing of the nervous system are relatively *indirect*. But every organic function is an interaction of intra-organic and extra-organic energies, either directly or indirectly. For life involves expenditure of energy and the energy expended can be replenished only as the activities performed succeed in making return drafts upon the environment—the only source of restoration of energy. (*Logic* 32, Dewey's emphasis).

Inquiry is—as are all aspects of experience—of the body *and* the mind, inextricable from the physical conditions that enable it, even if the importance of the physical environment is often masked by cultural constructs. The physical environment generates and restores all human potential. If we follow this line of thinking, the physical world powers everything we do, even writing and philosophizing. As Dewey plainly states:

> For the existential conditions which form the physical environment enter at every point into the constitution of socio-cultural phenomena. No individual person and no group *does* anything except in interaction with physical conditions. There are no consequences taking place, there are no social events that can be referred to the human factor exclusively. Let desires, skills, purposes, beliefs be what they will, what happens is the product of the interacting intervention of physical conditions like soil, sea, mountains, climate, tools and machines, in all their vast variety, with the human factor. The theoretical bearing of this consideration is that social phenomena cannot be understood except as there is prior understanding of physical conditions and the laws of their interactions. (*Logic* 485–86, Dewey's emphasis)

Phenomena can only be understood in the context of relations. The physical world always intervenes into the cultural world. "What exists, co-exists," in Dewey's words, "and no change can either occur or be determined in inquiry in isolation from the connection of an existence with co-existing conditions" (*Logic* 220). In other words, and this cannot be stressed enough, the social is embedded in the ecological.

Dewey goes on to claim that "If what is designated by such terms as doubt, belief, idea, conception, is to have any objective meaning, to say nothing of public verifiability, it must be located and described as behavior in which organism and environment act together, or *inter*-act" (*Logic* 40, Dewey's emphasis). Meaning-construction must be located, is an interaction between self and world. The human being is an organism that creates meaning, and "The organism is itself a part of the larger natural world and exists as organism only in active connection with its environment" (40). This active construction of meaning, or directed thought, always entails consequences: "There is no inquiry that does not involve the making of *some* change in environing conditions" (41, Dewey's emphasis). Inquiry is useful. It is a continuing process. Likewise, Dewey sees "objective meaning" as not static but dynamic, anticipating much poststructuralist thinking: "The attainment of settled beliefs is a progressive matter; there is no belief so settled as not to be exposed to further inquiry. It is the convergent and cumulative effect of continued inquiry that defines knowledge in its general meaning" (16). Meaning is determinable only in "a constellation of related meanings" (55), which are, in turn, emplaced in an environment, and, again, Dewey emphasizes interrelational processes among places, ideas, and things, intuits a pragmatist ecology—an ecology of progressive meaning.

For Dewey and for Williams, then, inquiry is located in both physical and cultural places. Again, as Casey argues, space is not something out of which places are carved; on the contrary, places enable the perception of space: space and time "arise from the experience of place itself" (37). Whatever we do, we are always positioned in a place or point of view. To perceive anything requires, of course, a body and its senses, and our bodies always reside in a place. Space, then, is perceivable only through the body and from a certain orientation point, from a place. Place becomes a central element in our development of a thinking, feeling self.[4] Along these lines, Gary Snyder writes: "A place on earth is a mosaic within larger mosaics—the land is all small places, all precise tiny realms replicating larger and smaller patterns. Children start out learning place by learning those little realms around the house, the settlement, and outward" (*Practice* 27). The world is a quiltwork of ecological and social patches in the context of which the self is formed.

All the writers considered in this study create an art that is emplaced. Their art is highly reliant on specific places, and those places function in the text not as settings, but as active participants in the work of art. In Karla Armbruster's estimation, the appearance of environmental agency separates the work of Williams from most other poststructuralist writing, which takes for granted a human subjectivity constructed by culture and history: "Williams sees nonhuman nature as one of the outside forces acting to structure subjectivity. In fact, she sees nature itself as the ultimate symbol of change and paradox rather than as the repository of unchanging, essential facts" (212). This destabilizes our received understanding of a work of art either as springing fully formed from the brain of the artist or as an artist's engagement with only the social milieu of his or her time. As we have seen with Dewey's understanding of inquiry, the work of art is an engagement with both the social and physical environments, with both subject and object. In the books investigated here, the role of the physical world is at least as important as the role of the social. In fact, they discourage thinking about the two categories as separate at all. This is not to say that I think that only work called "nature writing" can do this. I do think that these examples can help us understand more fully that, in a pragmatist ecology, the physical world, the social world, science, and work all participate in a work of art. Art can be ecological.

The work of art remains accessible over time, and its emplacement thickens the diachronicity of a literary text. Developing a strong argument for a diachronic understanding of literature, Wai Chee Dimock argues in "A Theory of Resonance" that a text "is not a finished product, a text is the incomplete expression of a finite language user; moving beyond that finite individual, it becomes a collective potentiality, a force of incipience commensurate with the incipience of humanity" (1064). Building on her argument, it is possible to introduce place into her theoretical equation. If a place is a central focus of a text, it seems reasonable to see the text commensurate also with the incipience of a place. The text, for Dimock, because of the continually shifting resonance of its context, has "a knack for vibrating on issues that matter" (1068). Environmental problems surely matter. In a later essay on the planetary reach of literature, Dimock writes, "Such resonance points to the importance of environmental 'background noise' as a generative force in literature"

("Planet" 179). I certainly agree; however, in all of Williams's work and all of the work examined in this study, that noise is brought much nearer the foreground. The intersection of culture and the natural environment itself generates literary art. Texts that open up questions about human relationships to the physical environment have a collective potential to perform inquiry and to intervene in dominant social discourse as Attridge has suggested texts can by changing conceptual domains. An addition of place into Dimock's notion of diachronicity loosens place from its traditional status of setting in a literary text. Places too can then be seen as participating in collective potentiality and demanding consideration as integral parts of finite selves and of literary texts that Dimock claims are crucial to democracy—they keep dialogue and inquiry open ("Resonance" 1068). Williams, too, is clearly concerned with the democratic valence of her work. Karl Zuelke argues that "The fusing of the speaking presence of the author with the surrounding landscape confers upon the landscape attributes of the speaking person.... Nature's person is established because it speaks. It follows that nature's speaking presence merits membership identity in a political realm, what we might term an 'ecopolitical space'" (240). Places become integral to aesthetics, ecology, and democracy. "In the Open Space of Democracy," writes Williams, "the health of the environment is seen as the wealth of our communities" (*Open Space* 8). If we recall Dewey's claims in *The Public and Its Problems,* we know too that "Democracy must begin at home, and its home is the local community," which, for Dewey, always finds itself emplaced though never static.

Williams in her turn emplaces her book and her self in the Utah landscape surrounding Great Salt Lake. Though installed in a natural field, a thing (a book) perceived alone is not a sufficient subject for inquiry. As Dimock has shown, the meaning of a text is never static. Neither are inquiry and its procedures; inquiry demands engagement with what Dewey calls a "situation," defined as "*not* a single object or event or set of objects and events. For we never experience nor form judgments about objects and events in isolation, but only in connection with a conceptual whole. This latter is what is called a 'situation'" (*Logic* 72, Dewey's emphasis). However, the situations that demand inquiry are not "whole"; rather, they are "indeterminate situations" that evoke questioning and are linked to the cultural and physical environment. Dewey writes: "We live and act in connection with the existing environment, not in connection with isolated objects, even though a singular thing may be crucially significant in deciding how to respond to total environment" (73). These situations are relational, rooted in environmental processes so that the very activity of inquiry is lodged in an ecosystem. Like places, situations exist in patterns and patches of relationships, and the resolution of singular problems aids in deciding how to respond to "total environment." In this view inquiry into particular situations embedded in particular places resonates throughout the environment. The relationship of writing to natural places remains an "indeterminate situation." Once place is untethered from the concept of setting and given a voice, meaningful inquiry into place and its textuality becomes possible and then becomes diachronically resonant.

It is important to pause here and keep in sight two of Dewey's key insights. First, he insists that "Only execution of existential operations directed by an idea in which ratiocination terminates can bring about the re-ordering of environing conditions required to produce a settled and unified situation" (*Logic* 121); in other words, inquiry is always directed toward practice that has real consequences in the existential world. And second, no situation is permanently "unified" or "settled," "inquiry is a *continuing* process in every field with which it is engaged" (16, Dewey's emphasis). All situations are involved in ongoing experience, to which aesthetic representation contributes. Williams attempts this reordering of situations embedded in a particular place and in a particular ecology. If we agree with Dimock that a text becomes collective potentiality and a force of incipience, then perhaps *Refuge* can be read as commensurate with the incipience of pragmatist ecological reasoning. And pragmatist ecological reasoning, as Dewey suggests, has democratic overtones: "A philosophy animated, be it unconsciously or consciously, by the strivings of men to achieve democracy will construe liberty as meaning a universe in which there is real uncertainty and contingency, a world which is not all in, and never will be, a world which in some respect is incomplete and in the making, and which in these respects may be made this way or that according as men judge, prize, love and labor. To such a philosophy any notion of a perfect or complete reality, finished, existing always the same without regard to the vicissitudes of time, will be abhorrent" ("Democracy" 50). Liberty is too important to be reified, to be turned into a slogan. It must be alive, and Dewey's warning thoughts here, published in 1919, seem strangely prescient. Although in the contemporary United States we seem stuck fast in the mire of a narrow utilitarian calculus, Dewey's vision is consistent with all that we know of the universe. Nothing remains the same, cultures and ecologies change, there is no balance of nature, only relative stabilities at best. In order to avoid becoming culturally and ecologically moribund, we must embrace and participate in the ever-changing world. In order for a democratic culture rooted in ecological principles to evolve, it must be always changeable, contingent. This is its power to keep healthy and breathing. Its lifebreath depends upon relations and interconnections.

Refuge is a logic of relations and interconnections. Williams's refuge is easily reached by car. It is only about an hour's drive from Salt Lake City, and Williams gives us exact directions how to get there, providing coordinates such as the Mormon Tabernacle and the Salt Lake City International Airport. The Bear River Refuge itself, though a home for wild creatures, is not a wilderness. According to Rachel Carson and Vanez Wilson in their 1950 pamphlet *Bear River: A National Wildlife Refuge*, its first recorded discovery by a European comes from Jim Bridger in 1824, and since then the river and marsh area have been repeatedly drained for irrigation. Between 1877 and 1900 more than two hundred thousand ducks were killed there annually for market. In 1928 the refuge was established and dams and canals were built to redistribute available water. Even though Williams's family, the refuge, the wildlife of the area, and the Great Salt Lake are intimately entwined in a specific history and ecology, the places in which their history and ecology are embedded have been officially (mis)perceived as empty space: "When the Atomic Energy Commission described the country north of the Nevada Test Site as 'virtually uninhabited desert terrain,' my family and the birds at Great Salt Lake were some of the 'virtual uninhabitants'" (*Refuge* 287). The places, ecologies, and the people in them were stripped of a political voice. Williams's home place is quite "real" and direly threatened by officially constructed perceptions of space. In her world, there is no empty space.

For Williams, space is the power that enables things to be connected. The most obvious clues to her understanding of space in these terms are her chapter titles, each of which names a type of bird, and the lake level. While we often bound space with the horizon, Williams is able to imagine, with the help of the birds, a deeper-reaching connection between space and experience. For example she thinks back to her childhood: "I . . . pondered the relationship between an ibis at Bear River and an ibis foraging on the banks of the Nile. In my young mind, it had something to do with the magic of birds, how they bridge cultures and continents with their wings, how they mediate between heaven and earth"

(*Refuge* 18). This notion of the bird's ability to bridge cultures and space has a venerable history, but interesting to this study are Carson and Wilson's similar comments on the birds of Bear River: "The importance of the Bear River Refuge is far more than local," and she goes on to say that "birds from Bear River have gone to Alaska, Canada, Mexico, Honduras, and Palmyra Island in the mid-Pacific" (11). Williams consistently teases out the relations between ecology and democracy, and one such nexus is in the Arctic: "The Arctic National Wildlife Refuge is the literal open space of democracy" (*Open Space* 59). Like William Cronon in his insistence that the tundra swan's ecosystem includes both the Arctic National Wildlife Refuge and Washington, D.C., both Carson and Williams perceive the birds as connecting ecosystems, cultures, and particular places. The birds gather experience from afar; they also mediate between self and other and, by implication, between self and something greater than the human, "heaven." Later, as an adult facing great personal loss, Williams again calls on the birds, and she, like Carson, is fascinated by the American flamingo: "How can hope be denied when there is always the possibility of an American flamingo or a roseate spoonbill floating down from the sky like pink rose petals?" (*Refuge* 90). The birds link human emotion—experience—with potentiating space that is anchored in a particular place, the Bear River Refuge. They also suggest that the importance of Bear River is not contained by the place. In a paradoxical way, a place is both local and universal: "a singular thing may be crucially significant in deciding how to respond to total environment" (*Logic* 73). As Dewey famously put it, "We are discovering that the locality is the only universal" ("Americanism" 13). One might say that through its textualization, place, given a voice, achieves a diachronic thickness.

While Williams imagines space as a power that enables an ecology, the particular ecology on which she feels able to focus is local. Again and again in Dewey's thought, "The local is the ultimate universal, and as near an absolute as exists" (*PP* 369). What I find so important for the future about Williams's understanding of ecology is that—although she indeed too often descends to male/female, culture/nature dichotomies—her ecological perception integrates history, love, family, sex, death, and nonhuman living and nonliving things. Therefore, *Refuge* is not only installed in a particular location but also "always embedded in a context of social relations" (Tallmadge, "Excursion" 202). Casting Williams as a writer with an eye toward consequences, Cassandra Kircher points out that these social relations extend outward into the world: "Williams's project in *Ref-*

uge [is] to reconceptualize her relationship with the world in language in order to change the world or, at least, in order to try to change the way that her readers perceive the relationship between humans and nature" (108). I would reiterate that it also seeks to give the physical world a voice. It is an important book that reflects a Deweyan attention to consequences and shows affinities with Deweyan logic. *Refuge* can make a strong contribution to our articulation of a pragmatist ecology.

As Mormons, Williams's family members are deeply rooted in the Utah landscape, and Williams adds that "The birds and I share a natural history. It is a matter of rootedness, of living inside a place for so long that the mind and imagination fuse" (*Refuge* 21).[5] Her place is not always a benign one. Williams imagines that "The pulse of Great Salt Lake, surging along Antelope Island's shores, becomes the force wearing against my mother's body" (64). At first the alignment seems a bit simplistic: the flooding lake threatens the bird refuge just as the cancer threatens William's mother. The island and the body of the mother are both places endangered by more or less natural occurrences, though the endangerment could also have been caused by government testing and human management of the landscape; thus, it becomes finally impossible to completely separate the cultural and the natural. Just the same, as Kircher points out, Williams breaks new ground with her alignment of the natural world and her actual mother, thereby reinscribing the family into environmental writing (104). Williams uses familial relations set in a particular place to inquire into the "indeterminate situation" of our current cultural understanding of the interrelationships among human and non-human communities.

To extend Kircher's argument, Williams attempts to create what might be called a familial ecology as her primary tool for performing inquiry. Again, inquiry happens within an "existential matrix," an environment. A problem, an "indeterminate situation," has a location. According to Dewey, "The immediate *locus* of the problem concerns . . . what kind of responses the organism shall make. It concerns the interaction of organic responses and environing conditions in their movement toward an existential issue" (*Logic* 111). That her mother's cancer may have been caused by nuclear testing upwind from her home, that Great Salt Lake is flooding the bird refuge, and that human beings evince a shocking disregard for the local environment all locate the specific problem around Williams's home place and help determine her response to the problem. Inquiry is generated through human interaction with an environment.

Because of her rootedness in the Utah landscape, Williams is in a perfect position to enact the type of interactive inquiry described by Dewey: "Organic interaction becomes inquiry when existential consequences are anticipated; when environing conditions are examined with reference to their potentialities; and when responsive activities are selected and ordered with reference to actualization of some of the potentialities, rather than others, in a final existential situation. Resolution of the indeterminate situation is active and operational" (111). Williams's primary response to the situation, of course, is *Refuge*. Within the book she makes certain choices, and one of these is her decision to rhetorically link the members of her family to the natural landscape in her attempt to present an ecology whose logic works toward a particular existential consequence: a revised cultural awareness that encourages people to live in a way more ecologically respectful toward the natural world and the human and nonhuman communities that inhabit it. Williams is concerned with the ongoing potential of human connection to very real places through the logic of a familial ecology.

Developing the images of this ecology, Williams claims that there are dunes hidden along the margins of Great Salt Lake that travelers, most of us motoring by on I-80, rarely manage to see: "[The] dunes . . . are the armatures of animals. Wind swirls around the sand and ribs appear. There is musculature in dunes. And they are female. Sensuous curves—the small of a woman's back. Breasts. Buttocks. Hips and pelvis. They are the natural shapes of Earth. Let me lie naked and disappear" (*Refuge* 109). Well, part of this is readily recognizable, even a bit clichéd. But there is a difference. Williams indeed figures the land as female, a strategy rightfully suspect to many. But it is not only a *human* female. The skeletal structure of animals is revealed by the work of wind upon sand, and ribs and muscles emerge. Only then are they revealed to be female, and only after that to be a woman's shape. For Williams the dunes are organic and productive, and the woman seems to evolve from this action upon the desert, from the touch of air upon land. She evolves from land to animal to human animal. But even more importantly, the woman can be seen as particular women, namely Williams, her mother, and her grandmother.

Williams herself has had two cysts removed from her breasts, both benign, and she worries about her own potential cancer. She writes about her grandmother's and her mother's physical beauty, undiminished by their mastectomy scars. At one point, as she stands looking at her own body in a mirror, her husband approaches from behind, and she whispers to him, "Hold my breasts" (*Refuge* 98). Because she aligns her own female body (and her potential cancer) with both the body of the sand dunes and her mother's dying body, this erotic gesture draws her male lover/husband deeply into the landscape and into the literal bodies of the family as well, thus extending the familial relationship well beyond the exclusively feminine.[6]

As her mother's cancer progresses, another link surfaces, the one between sex and death: "Death is no longer what I imagined it to be. Death is earthy like birth, like sex, full of smells and sounds and bodily fluids. It is a confluence of evanescence and flesh" (*Refuge* 219). The merging of her mother's body and natural spaces becomes even more clear: "I watch her skeleton push through skin, emerging bone by bone, rib by rib, until her vertebrae have become the ladder my fingers climb as I rub her back" (218). In a reversal of the process in which the human female body emerged from the sand, the dying female body recalls the dunes that "are the armatures of animals. Wind swirls around the sand and ribs appear" (109). By feeling her mother's rib cage emerge from her skin just as the ribs of female animals emerged from the dunes, Williams marks her realization of the cyclical nature of all life. As her mother declines, Williams charges this ancient awareness with the power that derives from familial love. In a move from familial to pragmatist ecology, she extends the family beyond the human: "I realize months afterward that my grief is much larger than I could ever have imagined. The headless snake without its rattles, the slaughtered birds, even the pumped lake and the flooded desert, become extensions of my family. Grief dares us to love once more" (252). Kircher points out that this passage dismantles patriarchy by depicting an "ever-expanding" family inclusive of all things (107). Catrin Gersdorf takes this idea a step further when she writes that for Williams, the Utah landscape "provides a whole archive of images which inform a language of sensual love and passion—be it for another human being, for an idea, or for the land—without drawing on metaphors of hierarchical gender relations, yet insisting on the metaphoric potential of the sensuous, lustful body" (175).[7] I agree on both counts, but even more happens here. Williams creates an "existential matrix" in which all things, physical and cultural, are part of an ecology, and all things become involved in her inquiry. She creates a pragmatist (eco)logic that privileges the ecotonal connections between the well-being of birds, sand, water, and lusty humans as integral parts of her inquiry.

An inclusive act of an imagination grounded in the ecology of a particular place allows Williams to love the desert and its creatures and landforms as family. Grief and despair transform into wonder and hope. Williams comes to see that "I am slowly, painfully discovering that my refuge is not found in my mother, my grandmother, or even the birds of Bear River. My refuge exists in my capacity to love. If I can learn to love death then I can begin to find refuge in change" (*Refuge* 178). Her willingness to embrace change in the form of death—not death in the abstract, but the "real" death of her mother—to see change in the form of a flooding lake, in the cyclical nature of human and nonhuman life, and to align that life with particular natural places enacts an (eco)logic that contributes to the practice of a pragmatist ecology. Williams's way lies through a specific place. The situations that drive our inquiry are closely related to particular places, to the bodies that inhabit those places, and to our representation of both. Williams says that "the most radical thing we can commit is to stay home" (qtd. in Jensen 322 and Siporin 101). Dewey seems to concur: "To remain and endure is a mode of action" (*Logic* 137). For both, our being in a place is our being along with a world, is our existence in a situational and situated ecology, in a logic that includes both cultural forms and physical environment.

Williams's method of inquiry accounts for the particularity of place and at the same time allows the significance of one place to resonate out into the culture. If attended to, it resonates out into the ecosystem. Again, the local is the only universal; the local powers the art of ecological knowing. Williams establishes a pragmatist (eco)logic that tacitly links her to all the other writers in this study and to all writers concerned with our relationships to the physical environment. In "A Eulogy for Edward Abbey," Williams writes, "Things are different now. That's why we're here. Change is growth, growth is life, and life is death. We are here to honor Ed, . . . to acknowledge family, tribe, and clan. And it has everything to do with love: loving each other, loving the land. This is a re-dedication of purpose and place" (202). Haines wrote an introduction to Muir's *Travels in Alaska*. Lopez honors Haines's poetry and dedicated a book to Carson. Carson wrote a guide to the place that is Williams's spiritual center. Williams wrote an *Audubon* article about Carson ("Spirit"). Williams and Lopez were deeply influenced by Steinbeck's writing. Steinbeck and Ricketts and Carson were reviewed together, and all three were fascinated by the *Coast Pilot*. In *The Sea around Us,* Carson listed Ricketts's *Between Pacific Tides* as essential reading. Perhaps John Muir and Steinbeck's character Joseph Wayne stretched out on the same rock. Their collective work stands along with the physical world. There is a diachronic significance in all this interrelationality, and Williams's understanding of family as a group of people bound by love and blood and of family as a collective potentiality of texts, people, and physical world that can all be brought to bear in performing inquiry is an indispensable model for a pragmatist ecology.

All of the authors in this study insist upon the importance of human relationships to particular places and particular ecologies. They all understand the actual in the light of the possible. Their writing gears us onto a physical world that participates in our shared experience; it is not simply the setting in which our experience happens. We are along with others, not beside them. Perception of our interrelatedness with human and nonhuman others can engender respect and wonder in the face of a bountiful world, and this respect, derived from wonder, "is a response worth cherishing and enhancing" (Greenblatt 53). This respect demands, of course, an ethical responsibility for the other, both human and nonhuman. Such a level of respect removes the human subject from a position of autonomous power, placing it in an ecosystemic context, thus revising the role of human culture in the natural community. This kind of respect implies an openness toward restraint. Respect, openness, and restraint are integral to pragmatist ecology and to democracy in its deepest sense. Dewey's words again resonate: "The clear consciousness of a communal life, in all its implications, constitutes the idea of democracy" (*PP* 328). This mindfulness also constitutes the idea of pragmatist ecology, a way of thinking and being centered in the ecotone in which ecology and democracy interpenetrate and in which our culture can resuscitate its potential to evolve into a more socially and environmentally just one. We remain a long way from achieving all this, but Derek Attridge, who writes that ethics makes impossible demands, also concedes that "ethically responsible acts occur every day" (30). Our art and inquiry may have the ability to alter perceptions in a wide cultural context, but only if we can nourish and nurture the art of knowing in the normal processes of living. As Dewey once again remarks: "One of the chief practical obstacles to the development of social inquiry is the existing division of social phenomena into a number of compartmentalized and supposedly independent noninteracting fields, as in the different provinces assigned, for example to economics, politics, jurisprudence, morals, anthropology, etc. . . . It is legitimate to suggest that there is an urgent need for breaking down these conceptual barriers so as to promote cross-fertilization of ideas, and greater scope, variety and flexibility of hypotheses" (*Logic* 501–2). In order for the art of knowing to flourish, and in order for all of our fields of inquiry to reach their potential, hardened borders must be broken down. Change is accelerated at the ecotone between ecology and democracy, between the natural and the cultural. Anything that impedes the vital flow of energy, dialogue, and exchange of information impedes the perception of relations that makes the aesthetic experience of this ecotone possible. Impediments to this flow then confound the conversion of despair for our environment and our democracy into hope, wonder, and activity.

I will close with an apt thought from Dewey: "Works of art are means by which we enter, through imagination and the emotions they evoke, into other forms of relationship and participation than our own" (*AE* 336). Aesthetic can be understood as ecological. Deweyan participatory politics, and a democratic people rooted in ecological values and in their local communities, can rejuvenate and sustain our already deeply diminished democracy and ecologies. The ecological understanding of experience—the pragmatist ecology—advocated by the writers in this study can help us respectfully participate every day with a world that is not, as we too often assume, entirely our own. We can live along with that world. We can be especially attentive to the fact that, as Williams warns, "When minds close, democracy begins to close" (*Open Space* 9). Finally, in the world in which we occur, a pragmatist ecology, following Dewey, insists that our responsibility lies foremost in understanding all problems, social and ecological, as remedial.

Notes

Introduction

1. Fesmire argues that "The moral of the arts is that everyday moral decisions can be as richly consummated as artistic productions" (128) and that moral responsibility toward the other can lead to "an ecological approach to the self-world relationship, of profound import for environmental policy" (87). In "John Dewey and the Moral Imagination," Alexander argues that "for Dewey, human beings are cultural creatures caught up in a dynamic interplay with their organic surroundings and their communicative systems" (383). In other words, human beings are entrenched in both social discourses and natural ecologies.

2. Shelley writes in *A Defence of Poetry:* "We want the creative faculty to imagine that which we know; we want the generous impulse to act that which we imagine; we want the poetry of life" (502–3). To act on what we imagine is fundamental to Dewey's pragmatism.

3. Allister engages the idea of an ecotone in much the same way that I am using it here. He refers in a helpful way to autobiography and environmental literature as hybrid genres. I also try to think about ecotones within texts themselves. But I could not agree more heartily with his phrasing, "Ecology intertwines with culture" (3).

4. In the "Editor's Note" to the winter 2003 issue of *Interdisciplinary Studies in Literature and Environment,* Scott Slovic introduces Jean Arnold's essay on the importance of Charles Darwin to American nature writing, "'From So Simple a Beginning': Evolutionary Origins of U.S. Nature Writing," as "picking up the thread of discussion about the relationship between literature and biology dangled in previous issues of *ISLE* by Glen Love, Jonathan Levin, and Bruce Clarke" (iv). Both Levin and Clarke argue convincingly for the importance of systems theory to our understanding of the world, our own embeddedness in it, and our neces-

sary difference from it. Levin takes Love to task for his "effort to draw a line between scientific method and socio-political values," which effort "suggests that our values can and should be derived in isolation from the processes by which we form (and reformulate) our knowledge about the world" (6). He goes on to invoke Dewey's consistent effort to employ scientific method as a critical tool that, among other things, explodes any notion of a value-neutral science (6). For Dewey, nothing created by human beings—least of all science—is value-neutral. Clarke, in his turn, refocuses Levin's discussion of Dewey onto "the mainstream specifics of the scientific history" of systems theory, although he too acknowledges that Dewey's work "suggests interesting analogues to systems-theoretical thinking" (160). Clarke urges the need to go beyond Anglo-American materials (153), and I agree, but I also feel that we can find good help in American thought, Dewey's in particular. See also Browne; Love, "Science"; and Love, *Practical Ecocriticism*. Love's is an extended argument for work that brings out the commonalities of the disciplines, particularly between literature and science. Love also insists that Darwin be awarded the attention in the context of literary study that Freud and Marx have.

5. I am indebted to Robert B. Westbrook's authoritative study *John Dewey and American Democracy*. Westbrook claims that Dewey "crafted a democratic philosophy of a depth and scope unparalleled in modern American thought" (552). In his concluding remarks, Westbrook offers the following advice: "Perhaps the most pertinent suggestion one could make to help stem the decay of democracy in concluding a book that I am painfully aware is likely to find an audience made up mostly of professors is to call on that audience not only for more Deweyan theory but also for more Deweyan practice" (552). See also his *Democratic Hope: Pragmatism and the Politics of Truth*.

6. In the acknowledgments to *The Open Space of Democracy*, Williams thanks Steven Rockefeller for pointing her back to Dewey's work, noting, "this has made an extraordinary difference" (n.p.).

7. *The Truth of Ecology* moves gracefully among current theoretical concerns, but, strangely, at the same time it often assumes a neoclassical tone in which Phillips sounds like Dr. Johnson sneering at the grounding of a metaphor. Regarding Emerson's famous transparent eyeball, Phillips writes: "Obviously Emerson's metaphor doesn't hold up from an ophthalmologic point of view: a 'transparent eye-ball' would be useless for the purpose of sight, since no light would be reflected by its retina (presumably, it wouldn't need to have a retina)" (200). Aside from seeming to insist on the same kind of realism he demands ecocritics eschew, this type of thing is not at all helpful; however, Phillips's final arguments that ecocriticism ought to take up the kind of messy nature that we can sink our teeth into, that we move from the pastoral into more troubled landscapes, seem to me vitally important.

8. See Dodson et al. and Wu and Loucks. Patch dynamics counters the idea that nature, left alone, tends toward equilibrium and balance. Nature itself is a

dynamic process of change. The ecosystem is a composite of ecological patches, all subject to disturbance. Wu and Loucks define "patch" as "a spatial unit differing from its surroundings in nature or appearance" (446). The larger ecosystem is a composite of these changing patches. The disturbances within patches can be natural or anthropogenic. This way of understanding ecosystem dynamics refuses to see nature as attaining a static balance. It denies a teleologically inflected view of the world and instead sees a natural world always in flux, always changing in response to both natural and cultural disturbance. This view of nature is strikingly similar to the way in which Dewey understood both the environment and the experience that occurs within it.

9. The sandbank passage has been often discussed. Sherman Paul writes that the passage is "an elaborate metaphor of the organic process of art and nature and self-reform, of the creative and shaping process of Idea, and the renewal that proceeds from the inside out" (*Shores* 346). The passage can also be seen as a renewal of experience. Gordon Boudreau writes that it is "a passage for Nature's deliverance that involved Thoreau as midwife in the process of converting to language what nature of herself could only struggle to articulate" (123). There is in this passage a notion of struggling conversation, an intense participation between nature and the human being. See also Anderson; Gura; Richardson; and West. David Robinson points out that in this passage "the human body, the organic world, and the earth itself are shown to be a single evolving entity" (94).

10. Robinson points out that Thoreau was indebted to Goethe's *Italienische Reise* for the figure of the leaf as form and for the importance of close attention to detail (43).

11. To avoid wandering off into metaphysics, in Dewey's work there exists always a fringe to precognitive experience. In other words, on the edge of our focus there are things barely or not quite perceived that can contribute to aesthetic experience. In *Experience and Nature,* Dewey credits William James with the idea of a focus and a fringe (236). Larry Hickman sheds some light on this difficult idea: "Dewey's radical empiricism also allows for the immediate experience of a 'beyond' in the sense that immediately experienced delight possesses sensible fringes. Hints, gaps, leads and clues are experienced on the fringes of focused experiences. In its non-cognitive phase, then, nature is the source of both felt delight and wider expectation" (65). These fringes, then, do not imply some metaphysical entity, but they acknowledge some aspect of the natural world or prior experience just outside our perception. This leads to expectation and to contingency in that these fringes always shift and change. Instability or contingency is to be celebrated, not dreaded. Contingency opens up the possibility of ongoing experience. It gives human beings the power to change given situations.

12. More and more work seems to support this balance. Philippon's *Conserving Words* is a powerful example, as is Peterson's *Being Human.*

13. The quoted passage is from a talk Dewey gave to the National Education

Association in 1937. Published in essay form as "Democracy and Educational Administration," it can be found in Dewey's *Later Works* 11: 217–25. The quoted passage is on page 225.

Chapter 1

1. For Muir's biography, see Holmes; Wilkins; Stanley; F. Turner; M. P. Cohen; Fox; Wolfe, *Son of the Wilderness;* and Badè. Much of Muir's own work, garnered, like *My First Summer in the Sierra*, from old journals and notes, is autobiographical, especially *The Story of My Boyhood and Youth* and the posthumously published *A Thousand-Mile Walk to the Gulf*.

2. See also Branch, "Muir's *My First Summer*," esp. 141–43. Branch argues that "It is useful, in fact, to think of the book as being co-authored by two Muirs: the 31-year-old Muir encountering the high sierra for the first time in the summer of 1869, and the 73-year-old Muir using his literary skills to convey the importance of that encounter in a book published 42 years later" (142).

3. See Philippon's excellent chapter on Muir, "Our National Parks: John Muir and the Sierra Club." Philippon also examines Theodore Roosevelt and the Boone and Crocket Club, Mabel Wright and the Audubon Society, Aldo Leopold and the Wilderness Society, and Edward Abbey and Earth First!

4. See especially Nash's chapter "Hetch Hetchy" (*Wilderness* 161–81) for a detailed history of the Hetch Hetchy controversy and its importance in bringing to national attention efforts to preserve wilderness. Nash writes: "Previously most Americans had not felt compelled to rationalize the conquest of wild country in this manner. For three centuries they had chosen civilization without any hesitation. By 1913 they were no longer so sure" (181).

5. Worster writes: "Indeed, the conservation program that emerged under Pinchot's leadership in the early years of the twentieth century paid little attention to ecological complications. It was primarily a program aimed at maximizing the productivity of those resources in which man had a clear, direct and immediate interest" (269). Pinchot and Muir split most vehemently over forest reserves and the Hetch Hetchy controversy. See M. P. Cohen 325–38; Nash, *Wilderness* 133–40; and Runte 78–79.

6. On shifting patterns of perception and representation in the nineteenth century see Hönnighausen. On this particular move from a Romantic to Victorian location of meaning see 123–28.

7. On the book of nature see St. Armand; on Muir and Agassiz and Agassiz's concept of nature as a book see M. P. Cohen 104–11, 176–77.

8. For Dewey, the concept of unity is not necessarily static. Unity is not a goal in itself, but it too belongs to experiential process. According to Richard Shusterman, Dewey "repeatedly insists that the unity of aesthetic experience is not a closed and permanent haven in which we can rest at length in satisfied contemplation. It is rather a moving, fragile, and vanishing event, briefly savored in an experiential flux rife with energies of tension and disorder which it momentarily

masters. It is a developing process which, in culmination, deconstructively dissolves into the flow of consequent experience, pushing us forward into the unknown and toward the challenge of fashioning new aesthetic experience, a new moving and momentary unity from the debris and resistance of past experiences and present environing factors" (32).

9. See especially Solnit 250–67. Scheese discusses landscape art throughout his first chapter, "Overview."

10. See also Buell's comments on this passage in *The Environmental Imagination* 157. Evernden's questions call for a relinquishment of the autonomous self.

11. When Muir was a boy in Scotland, his father tried with little success to restrict him to the enclosed backyard, and years later, in perhaps his most traumatic moment, Muir almost suffocated in the bottom of a well his father forced him to dig through solid sandstone. At the time of the incident, he had already chipped his way eighty feet down, and after two days' recovery time his father lowered him back down into the well. See *The Story of My Boyhood and Youth*, 2, 232.

12. See especially Fleck's chapter 2, "John Muir's Homage to Henry David Thoreau." At the back of each of his volumes of Thoreau's writings, Muir penciled in his own index. He knew Thoreau's work intimately.

13. Muir, "Twenty Hill Hollow" 80. The opening of Thoreau's essay is worth quoting here because it clearly had a powerful effect on Muir's life: "I wish to speak a word for Nature, for absolute freedom and wildness, as contrasted with a freedom and culture merely civil—to regard man as an inhabitant, or a part and parcel of Nature, rather than a member of society. I wish to make an extreme statement, if so I may make an emphatic one, for there are enough champions of civilization: the minister and the school committee and every one of you will take care of that" ("Walking" 205). Surely Muir had had enough of ministers and the stern Calvinism of his father. He, like Thoreau, wished to speak for wildness, and believed in that other famous statement from the same essay, "in Wildness is the preservation of the world" (224). See also Stewart 121.

14. Muir had of course felt this kind of tension before. The famous last lines of his *Story of My Boyhood and Youth* read, "But I was only leaving one University for another, the Wisconsin University for the University of the Wilderness" (287). The lines also appear in "Out of the Wilderness" 277.

15. On the meeting of Muir and Emerson see Branch, "'Angel Guiding Gently'"; and Hansen.

16. See Hönnighausen 106–7. See also Walls 42–44 (on Lyell) and 194–99 (on Darwin).

17. For a detailed examination of Muir's use of Ruskin and of the marginalia in Muir's copy of *Modern Painters*, see Gifford, "Muir's Ruskin." Gifford writes: "What both Muir and Ruskin addressed from different perspectives, Ruskin essentially anthropocentrically and Muir essentially biocentrically, was the dilemma of human presence, influence and responsibility for the earth" (144). See also Hadley 492–501.

18. Coleridge talks about the esemplastic imagination in chapter 13 of *Bio-

graphia Literaria. Dewey talks often about Coleridge, and he discusses the esemplastic imagination in *Art as Experience* (272). Dewey finds in Coleridge's discussion that "an imaginative experience is what happens when varied materials of sense quality, emotion, and meaning come together in a union that marks a new birth in the world."

19. In a wonderful phrase, O'Grady calls Muir "a peripatetic lover of storms and cascades" (48).

20. Philippon makes the point that early in his career Muir ordered his morality vertically also. Philippon sees this as a reversal of Muir's father's Calvinism, which sought to bring light to the darkness of the wilderness. See especially 130–32.

Chapter 2

1. The best discussion of *To a God Unknown* is Robert DeMott's "'Writing My Country': Making *To a God Unknown*" (*Steinbeck's Typewriter* 108–45).

2. See Timmerman 315–16. Timmerman argues that Steinbeck in *To a God Unknown* voices an early, relatively unformed environmental ethic. Because Joseph Wayne fails to live in harmony with the land until his death, the ethic fails. In Timmerman's estimation, any ethic that results in a corpse is not really an ethic at all; ethics are for "living people" interacting with "living environments."

3. See Cassuto. Cassuto looks at the historical context of *The Grapes of Wrath* and argues that the confluence of national myth, bad farming practices, and disregard of normal drought cycles led to the ecological devastation in the plains states in the 1930s. Also, a refusal of new thinking contributed to the disaster. Cassuto writes: "It is precisely this sort of stubborn adherence to traditional values while implementing ecologically pernicious agricultural methods that brought on the 'dirty thirties'" (60). Steinbeck and Ricketts's deep conviction that a new way of thinking was necessary permeates *Sea of Cortez*. Surely, writing *The Grapes of Wrath* spurred Steinbeck on to new thinking.

4. See also DeMott, *Steinbeck's Reading* 155. During the time he was writing *East of Eden*, Steinbeck also read William James's *Varieties of Religious Experience* and Dewey's *A Common Faith*.

5. About a possible smear campaign by the Associated Farmers of California, see Fensch 20–21. See also Benson 419–26 for a detailed discussion, including accusations that Steinbeck was part of a Jewish plot. Elaine Steinbeck and Robert Wallstein's collection *Steinbeck: A Life in Letters* contains Steinbeck's reply to Rev. L. M. Birkhead, the head of an anti-Nazi group trying to counter anti-Semitic propaganda in the United States, who wrote to ask Steinbeck if he was Jewish. Steinbeck replied that he was not, but "Those who wish for one reason or another to believe me Jewish will go on believing it while men of good will and good intelligence won't care one way or another. I can prove these things of course—but when I shall have to—the American democracy will have disappeared" (203–4).

Also, in April 1940, in a letter to Mrs. Franklin D. Roosevelt, who had visited the California migrant camps and reported what she saw to the *New York Times,* Steinbeck wrote: "Meanwhile—may I thank you for your words. I have been called a liar so constantly that sometimes I wonder whether I may not have dreamed the things I saw and heard in the period of my research" (202). Little wonder Steinbeck wanted to sail away.

6. In 1899, Ritter joined a group of naturalists that included John Muir on an expedition to Alaska sponsored by E. H. Harriman. Ritter worked all hours taking marine samples, and Muir writes in his journal for July 4, "Charles Keeler read a poem, Ritter and Fernow danced jigs." Wolfe, *Son of the Wilderness,* 281; *John of the Mountains,* 401, 412.

7. Clearly, Steinbeck and Ricketts felt that the generation of new questions aimed toward better understanding the living community of the Sea of Cortez outweighed the life of, say, an individual heliaster. Of course, as Railsback's book proves, Darwin was a major force in all of Steinbeck's work, especially *Sea of Cortez.* Stanley Brodwin sets *The Log from the Sea of Cortez* in a tradition of scientific travel writing, and a name I have not mentioned, Alexander von Humboldt, is a key figure in his essay. Humboldt also loomed large in Muir's development. See Holmes 130–34. See Walls (esp. chapters 3 and 4) for Thoreau's debt to Humboldt.

8. I draw here on Railsback's discussion in *Parallel Expeditions,* 1–9.

9. The two studies are Ventor et al., "The Sequence of the Human Genome," and Collins et al., "The Human Genome." The first is privately sponsored by Celera Genomics. The second is publicly sponsored, and although it was led by Collins, more than twenty-five hundred authors from twenty laboratories participated in the release of this project's findings.

10. John Steinbeck, *Sea of Cortez,* Autograph manuscript unsigned of the first draft, MA 3652, 1940, Pierpont Morgan Library, New York City.

11. Peattie has this to say about Ricketts's influence on Steinbeck: "[Ricketts] has given his ebullient friend [Steinbeck] a biological philosophy—the best kind a man can have, to my way of thinking, the most realistic, and yet the kindest. Mr. Steinbeck appears to have discovered it only yesterday, in its full implications and deep perspectives, and he is wildly excited about it. He runs up and down the echoing corridors, paced by Aristotle, Lucretius, Goethe, and the Huxley boys, each in their day, turning cartwheels of mental liberation from final causes, shouting at the marble walls, enchanted by the echoes that they throw back of his own voice, and stopping, sometimes, to whistle through his teeth at the scholastic dignitaries" (6). Although his review is mostly positive, Peattie sees *Sea of Cortez* as the death knell for Baja California (masses of tourists will follow), and he objects to a tendency to moralize that he hears in the book. I do not think he takes Steinbeck quite seriously enough. Peattie's review is one of a pair in this issue under the heading "Of and about the Sea." The other is a review of Rachel Carson's *Under the Sea Wind* by William Beebe. On a note of related interest, when Carson

submitted the first chapter of *Under the Sea Wind,* "Flood Tide," to *Atlantic* magazine, it was rejected because the editor had already committed to a series of nature essays by Donald Culross Peattie (see Lear 98).

12. John Steinbeck to Elizabeth Otis, 26 March 1940, *Life in Letters* 201. This letter was written onboard the *Western Flyer.*

13. John Steinbeck to Carlton A. Sheffield, 13 November 1939, *Life in Letters* 194.

14. Although Abram is primarily engaged with phenomenology and Maurice Merleau-Ponty, in the context of relations there are many crossings between pragmatism and phenomenology. For instance, in *Phenomenology of Perception,* Merleau-Ponty writes: "To be a consciousness or rather *to be an experience* is to hold inner communication with the world, the body and other people, to be with them instead of being beside them" (96, Merleau-Ponty's emphasis). The idea of experience as participation, of being "with" instead of "beside," resonates with Dewey's thought, but I think Dewey would insist less upon "inner" communication and more upon mental and physical interrelations with the world. See Kestenbaum, *The Phenomenological Sense of John Dewey.* See also Rosenthal and Bourgeois.

15. Williams chose *To a God Unknown* as the most important novel in her experience ("The Outside Canon").

16. Perhaps an analogy to certain territorial fish can help illustrate this idea. Evernden explains that the stickleback's behavior changes relative to the shape of its territory: "the boundary of what the fish considers to be himself has expanded to the dimensions of the territory. He regards himself as being the size of the territory, no longer an organism bounded by skin but a organism-plus-environment bounded by an invisible integument" (44). We do not necessarily live inside our skin.

17. See Bakhtin, *Dialogic Imagination* 284 and *Speech Genres* 93; and Bhabha 149.

18. Compare Steinbeck's perception here to Thoreau's in "The Ponds" chapter of *Walden:* "Sometimes, after staying in a village parlor till the family had all retired, I have returned to the woods, and, partly with a view to the next day's dinner, spent the hours of midnight fishing from a boat by moonlight, serenaded by owls and foxes, and hearing, from time to time, the creaking note of some unknown bird close at hand. These experiences were very memorable and valuable to me, anchored in forty feet of water, and twenty or thirty rods from the shore, surrounded sometimes by thousands of small perch and shiners, dimpling the surface with their tails in the moonlight, and communicating by a long flaxen line with mysterious nocturnal fishes which had their dwelling forty feet below, or sometimes dragging sixty feet of line about the pond as I drifted in the gentle night breeze, now and then feeling a slight vibration along it, indicative of some life prowling about its extremity, of dull uncertain blundering purpose there, and slow to make up its mind. At length you slowly raise, pulling hand over hand, some horned pout squeaking and squirming to the upper air. It was very queer, especially in dark nights, when your thoughts had wandered to vast and cosmogonal

themes in other spheres, to feel this faint jerk, which came to interrupt your dreams and link you to Nature again. It seemed as if I might next cast my line upward into the air, as well as downward into this element, which was scarcely more dense. Thus I caught two fishes as it were with one hook" (117).

Chapter 3

1. For instance, the first literary critic to examine Carson's work was Carol Gartner, who writes in *Rachel Carson:* "Although each of Carson's books is a separate literary work, they form a developmental sequence reflecting her growing sense of ecological relationships and increasing foreboding at the effect of our destruction of the natural world" (3).

2. Plumwood writes: "Ecofeminism has particularly stressed that the treatment of nature and of women as inferior has supported and 'naturalized' not only the hierarchy of male to female but the inferiorization of many other groups of humans seen as more closely identified with nature. It has been used to justify for example the supposed inferiority of black races or indigenes (conceived as more animal), the supposed inferiority of 'uncivilized' or 'primitive' cultures, and the supposed superiority of master to slave, boss to employee, mental to manual worker. For Western society, which has particularly employed a genderized concept of nature as a way of imposing a hierarchical order on the world, feminization and naturalization have been crucial and connected strands supporting pervasive human relations of inequality and domination both within Western society and between Western society and non-Western societies. The interwoven dualisms of Western culture, of human/nature, mind/body, male/female, reason (civilization)/nature, have been involved here to create a logic of interwoven oppression consisting of many strands coming together" (211).

3. See Stewart 163 and 252 n. 6. See especially Lear 448–55: after the publication of *Silent Spring,* Carson appeared on the news show *CBS Reports* and was interviewed by Eric Sevareid. Some of the largest sponsors canceled their support of the program, among them Lehn and Fink (the makers of Lysol), Standard Brands, and Ralston Purina. For Carson, "most gratifying of all, the day after *CBS Reports* aired, Senator Hubert Humphrey (D-Minnesota) announced that he had asked Senator Abraham Ribicoff (D-Connecticut) to conduct a broad-ranging congressional review of environmental hazards, including pesticides. Ribicoff would chair a subcommittee of the Senate Government Operations Committee to hold hearings for that purpose. Two weeks later Carson accepted an invitation to testify before the Ribicoff committee" (Lear 450–51). During the hearings, Senator Ernest Gruening (D-Alaska) compared *Silent Spring* to *Uncle Tom's Cabin.* Nash writes that "One-hundred and ten years after *Uncle Tom's Cabin* Rachel Carson wrote another book that exploded against traditional American assumptions. It argued that all life-forms, even insects, were not commodities but deserved ethical consideration" (*Rights* 78).

4. Interestingly, *Silent Spring* appeared in the same year as Thomas Kuhn's

The Structure of Scientific Revolutions, and both of these books undercut the authority of traditional reductionist science.

5. However, as Glotfelty notes, U.S. companies continue to produce and export massive quantities of DDT ("Rachel Carson" 165).

6. Predictably, Phillips takes Worster to task for expecting ecology to provide a "moral compass" (49). I agree that Worster could be more open to the theoretical shifts within the science of ecology, but, certainly, ecology provided Carson with a moral compass—and ethical sonar and aesthetic radar. Perhaps that is why Phillips does not mention her in his book.

7. Carson also wrote a number of pamphlets for a Fish and Wildlife Service series called *Conservation in Action.* In an interesting crossing, pamphlet 8, coauthored by Carson and Vanez T. Wilson and illustrated by Bob Hines, is about the Bear River Migratory Bird Refuge, which forms the nucleus of Terry Tempest Williams's book *Refuge,* discussed in the conclusion of this study. In the spirit of Carson, Williams condemns nuclear testing in the American West.

8. "The Real World around Us" was originally a talk given to the Sorority of Women Journalists in 1954.

9. Gartner reaches the same conclusion, calling *Under the Sea Wind* Carson's "most successful literary work" (*Rachel Carson* 29). It was also Carson's favorite.

10. Carson's posthumously published *The Sense of Wonder* is a guidebook to teaching a child (and an adult) to wonder. Part of the element of wonder is the recognition that there is something beyond the human: "there is symbolic as well as actual beauty in the migration of birds, the ebb and flow of the tides, the folded bud ready for the spring. There is something infinitely healing in the repeated refrains of nature—the assurance that dawn comes after night and spring after the winter" (88–89).

11. Although she does, of course, use fable in a rhetorically powerful way at the beginning of *Silent Spring.*

12. Lear cites (505 nn. 31, 32) a letter from Carson to Hendrik van Loon dated 5 February 1938. The letter is in the Rachel Carson Papers in the Beinecke Rare Book and Manuscript Library, Yale University.

13. In "Memo to Mrs. Eales on *Under the Sea Wind,*" Carson writes: "As far as possible, I wanted my readers to feel that they were, for a time, actually living the lives of sea creatures. To bring this about I had first, of course, to think myself into the role of an animal that lives in the sea. I had to forget a lot of human conceptions" (*Lost Woods* 55–56).

14. See Gartner, *Rachel Carson* 29, 36–38. Gartner takes up the relationship between science and literature again in "When Science Becomes Literary Art."

15. Glotfelty explores Carson's use of cold war rhetoric in "Cold War, *Silent Spring.*" Carson turned the rhetoric of war used by the pesticide makers back on themselves, and "the lexicon of war continues to pervade environmental discourse" (159).

16. Among Carson's "Suggestions for Further Reading" appended to the text

of *The Sea around Us* is Edward Ricketts and Jack Calvin's *Between Pacific Tides*. Steinbeck wrote the preface to the third edition of *Between Pacific Tides*.

17. Incursions of shipborne rats into Alaskan ecosystems continues to pose a serious threat, especially to bird populations. See, for example, the Union of Concerned Scientists website: http://www.ucsusa.org/news/press_release/new-report-finds-invasive-species-degrading-alaskas-ecosystems.html.

18. Coleridge, *Biographia Literaria,* esp. chapters 13 and 14. Art remains to a degree subordinate to nature, but poetry, or art, is entirely of the imagination. Although admittedly influenced by Romantic ideas, Dewey and Carson insist that aesthetic experience results from the interrelationship of organism and environment. Where a typical Romantic formula privileges the imagination over the natural world, Carson, and an environmental aesthetics in general, attempts to better balance the equation.

19. Williams certainly learned from Carson's understanding of ecology. See her "The Spirit of Rachel Carson."

20. Barry Lopez writes in "Landscape and Narrative" that "The interior landscape responds to the character and subtlety of an exterior landscape; the shape of the individual mind is affected by land as it is by genes" (*Crossing Open Ground* 65).

21. "A dull man seems to be a dull man no matter what his field, and of course it is the right of a dull scientist to protect himself with feathers and robes, emblems and degrees, as do other dull men who are potentates and grand imperial rulers of lodges of dull men" (*Cortez* 73).

22. As Peterson points out, the idea of relationality is also at the center of most feminist thought. See chapter 6, "Relationships, Stories, and Feminist Ethics," in *Being Human.*

23. By the time she wrote *Silent Spring,* Carson had come to realize the nightmarish side to the principle of interrelation. The introduction of pesticides into ecosystems has ramifications all along the food chain. This is another example of something we know theoretically but seem unable or unwilling to fully engage in practice.

24. Taylor argues that Chaloupka's claims for Dewey as an environmental thinker or as a source for environmental thinking are overstated and misleading. He insists that Dewey's work is profoundly human centered and that, in Dewey's thought, nature is for the use of human beings. I guess this comes back to how we want to think about the term "use." Surely, Dewey did not see the same environmental problems we see today, and no one is claiming that of him. But his idea that human experience is rooted in nature seems to me undeniable. Humans and the nonhuman world are both important because they participate in shared experience. I do not claim that Dewey ever thought that nonhuman things had inherent rights. The nonhuman, again, must be cared for and respected because of its contribution to experience. Value resides in the quality of the process. We cannot think of anything, much less grant a thing rights, somehow standing outside

of our own, human experience. Further, Taylor writes that "There is no sensitivity in his work to the possibility that scarcity of resources might become a significant problem for our society" (184). However, a passage I quoted in chapter 2 seems pretty convincing evidence that Dewey did indeed care plenty about distribution and scarcity of resources: "Conservation of not only the public domain but restoration of worn-out land to fertility, the combating of floods and erosion which have reduced vast portions of our national heritage to something like a desert, are the penalties we have to pay for past indulgence in an orgy of so-called economic liberty. Without abundant store of natural resources, equal liberty for all is out of the question. Only those already in possession will enjoy it. Not merely a modification but a reversal of our traditional policies of waste and destruction is necessary if genuine freedom of opportunity is to be achieved ("Freedom" 251). Taylor wants us to pay attention to Dewey's ideas on democracy first, but I think that in a pragmatist ecology based on Dewey's thinking it is far more helpful and more in the spirit of Dewey's philosophy to think that ecology and democracy might be part of a continuum, not two entirely separate things.

25. See McDonald on continuity, especially his chapter "Dewey's Naturalism." The human organism is not separate from nature but is part of nature, is continuous with it. The human is always in an environment, not separate from it.

26. In his review of *Sea of Cortez*, Lyman takes Steinbeck and Ricketts to task for failing to take into consideration the chemical and physical characteristics of the seawater.

27. See also Buell's excellent chapter in *The Environmental Imagination* on "Environmental Apocalypticism," 280–308. Buell compares Carson's *Silent Spring* with Leslie Marmon Silko's *Ceremony*. World War II and the dire potential of technological destruction pervade both otherwise very different books. On a related matter, although Carson had some faith in the biological manipulation of insects, Buell suggests that she would have been less enthusiastic had she confronted current developments in genetic engineering (521 n. 52).

28. Some of the opposition to Carson at the time of her writing is certainly still recognizable. McCay writes that "Perhaps it is Carson's willingness to accept nature's incomprehensible work that left her open to the challenge and the insinuations of the chemical companies when she published *Silent Spring*. A person who accepted the possibility of mystery could not be a 'scientist.' Admitting there were some things in nature we could not know, could not control, limited man's dominion. Thus Carson, because she cautioned humility in the face of nature, was considered unscientific. Humility is also a feminine virtue, not one likely to be valued by the conglomerate of chemical companies arrayed against her" (108). In what seems an outburst of insanity, ultraconservative columnist Thomas Sowell in his syndicated column for 9 June 2001 writes: "Rachel Carson and the environmentalists she inspired have succeeded in getting DDT banned in country after country, for which they have received the accolades of many, not least their own accolades. But in terms of the actual consequences of that crusade, there has not

been a mass murderer executed in the past half-century who has been responsible for as many deaths of human beings as the sainted Rachel Carson" (A12).

29. Brooks, Carson's close friend and publisher, writes that "No American authors have written more eloquently about nature than Henry Thoreau and Rachel Carson" (*Speaking for Nature* xii).

30. Carson was posthumously awarded the Presidential Medal of Freedom by Jimmy Carter in 1980. The citation reads: "Never silent herself in the face of destructive trends, Rachel Carson fed a spring of awareness across America and beyond. A biologist with a gentle, clear voice, she welcomed her audiences to her love of the sea, while with an equally clear determined voice she warned Americans of the dangers human beings themselves pose for their own environment. Always concerned, always eloquent, she created a tide of environmental consciousness that has not ebbed" (qtd. in Gartner, *Rachel Carson* 28 and in Glotfelty, "Rachel Carson" 169).

31. On Carson's rhetoric see Glotfelty, "Rachel Carson," esp. 161–63; see also her "Cold War, *Silent Spring*."

32. On the rhetorical use of "home" and "family," see Norwood, "The Nature of Knowing."

33. The letter was written by an H. Davidson to the *New Yorker*.

Chapter 4

1. In the pamphlet *John Haines,* Peter Wild makes much of Haines's indebtedness to Bly, but he largely fails to understand the distinction Marty Cohen makes between the denatured aspects of mainstream poetry and the kind of natured experience Haines incorporates in his best poetry. Wild falls far short of coming to terms with the ecology of Haines's work. The pamphlet was written before the publication of *The Stars, the Snow, the Fire,* and at its conclusion Wild writes: "One can only hope that as he enters his seventh decade, he not only will keep goading American poets toward better work, but also that for the sake of his readers he will complete the full-length autobiography already foreshadowed in the latter part of *Living Off the Country*" (48). That autobiographical work is, of course, *The Stars, the Snow, the Fire,* and in a 1990 review, Wild writes: "As a painter and one of our country's foremost poets, Haines has done more than his share to help preserve the nature he loves. Many people have gone into the wilderness and been inspired; only a handful have emerged with the enviable ability of a John Haines to enrich us with first-rate stories about how—and how deeply—they lived" (94). Reading *The Stars, the Snow, the Fire* (and perhaps *New Poems*) seems to have influenced Wild's critical estimation of Haines's contribution as a whole.

2. It seems he will remain so. Valerie Trueblood writes: "John Haines took an inspired course, in which the way he undertook to live made room for, tested, and determined his art. The Alaska that formed him is half gone. Eco-tourists sneak up on the animals that were danger, prey, mystery, and nightmare to him.

Admirers of his poetry can only hope to make the journey out of the new century's sentimental and mechanical cultures and find a North of their own. Meanwhile Haines is blessed to have become the poet who, appalled witness to a century, could still make us see the Green Man with shedding dogs attending him and a maggot singing in his beard as more than a conceit, and the one poet in America who could begin a poem: 'Slowly, without sun, the day sinks toward the close of December. It is minus sixty degrees'" (50). Haines here is clearly tied to the physical landscape of Alaska, and he is also enmeshed in the political role of the common person and the physical world in our threatened democracy: "politics for Haines has two meanings, the huckster one with its deceptions and the real one as lived by people in society. This second politics has nothing to do with a platform; it has to do with the common life. Its connection with government is that the citizen, the *polites,* gets his living and often the manner of his death under the auspices of some government. Politics thought of in this way is the form by which the fight to live communicates itself, among people of many incompatible needs. (And nowadays the term extends to the voice in art and debate given to beleaguered nature.)" (47). Trueblood is right on the mark with her understanding of Haines's interest in the common life and his attempt to endow nature with a political voice through his art. Nature, in a Deweyan sense, participates in his art. Steven B. Rogers has recently edited a tribute to Haines: *A Gradual Twilight: An Appreciation of John Haines.*

3. I suggest that this is more than a literary tactic of personification. We think of the physical world as a living entity more than we care to admit, and it would repay further study to examine the way landforms are described in scientific texts. For instance, the following passage in a text on geomorphology certainly perceives streams as having some level of agency: "The process of headward erosion of a stream until it undercuts and captures another stream is known as abstraction. Streams in the headwaters of a drainage system have an advantage over competing streams and are more likely to capture them if they have the following characteristics" (Easterbrook 149).

4. Altieri writes: "Art is literally empowerment: the making available of exemplifications that enable us to look at ourselves, as we encounter different sites of being and modes of inhabiting them. And the direct testimony provided by such examples promises to free its audience from its dependency on the entire apparatus of representation, and from the positing of interpretations trapped within the narrow expectations cultivated by the ideology of 'aesthetic emotion.' There we find ideals that may generate significant social change" (56). I am interested in this move from "aesthetic emotion." Perhaps one way to think of this is as a move toward Deweyan aesthetic experience.

5. In his final chapter, "Writing Off the Self," Poirier introduces the possibility that the attempt to dissolve the self and the resultant awareness of a prior world can result in moments of wonder and beauty (202). Such moments are also central pivots in *The Stars, the Snow, the Fire.*

6. Haines's connection between work and land echo Steinbeck's in *The Grapes*

of Wrath: "Then the corrugated iron doors are closed and the tractor man drives home to town, perhaps twenty miles away, and he need not come back for weeks or months, for the tractor is dead. And this is easy and efficient. So easy that the wonder goes out of work, so efficient that the wonder goes out of the land and the working of it, and with the wonder the deep understanding and the relation. And in the tractor man there grows the contempt that comes only to a stranger who has little understanding and no relation" (157). This passage can be read as a metaphor for our culture's mechanistic relationship to the land more generally.

7. Aside from Barry Lopez's *Arctic Dreams*, which I discuss in the following chapter, two other texts immediately come to mind that mark historical moments in the changing aspects of life in the North: Edward Hoagland's *Notes from the Century Before: A Journal from British Columbia* and John McPhee's *Coming into the Country*.

Chapter 5

1. Lopez has spoken out on this idea before, perhaps most explicitly in *The Rediscovery of North America*. After cataloging the atrocities of the Spanish, he continues: "I single out these episodes of depravity not so much to indict the Spanish as to make two points. First, this incursion, this harmful road into the 'New World,' quickly became a ruthless, angry search for wealth. It set a tone in the Americas. The quest for personal possessions was to be, from the outset, a series of raids, irresponsible and criminal, a spree, in which an end to it—the slaves, the timber, the pearls, the fur, the precious ores, and, later, arable land, coal, oil, and iron ore—was never visible, in which an end had no meaning. The assumption of an imperial right conferred by God, sanctioned by the state, and enforced by a militia; the assumption of unquestioned superiority over a resident people, based not on morality but on race and cultural comparison—or, let me say it plainly, on ignorance, on a fundamental illiteracy—the assumption that one is *due* wealth in North America, reverberates in the journals of people on the Oregon Trail, in the public speeches of nineteenth-century industrialists, and in twentieth-century politics. You can hear it today in the rhetoric of timber barons in my home state of Oregon, standing before the last of the old-growth forest, irritated that anyone is saying '*enough* . . . , it is enough.' What Columbus began, then, what Pizzaro and Cortés and Coronado perpetuated, is not isolated in the past. We see a continuance in the present of this brutal, avaricious behavior, a profound abuse of the place during the course of centuries of demand for material wealth. We need only look for verification at the acid-burned forests of New Hampshire, at the cauterized soils of Iowa, or at the collapse of the San Joaquin Valley into caverns emptied of their fossil waters" (9–11). His second point is that we need not be bound by the past.

2. Lynn White Jr. traces the ecological crisis back to traditional Christianity. On the potential contribution of Christianity to environmental ethics, see Peterson; Oelschlaeger; McFague; and Hefner.

3. Shelley thought so too: "The cultivation of those sciences which have enlarged the limits of the empire of man over the external world, has, for want of the poetical faculty, proportionally circumscribed those of the internal world; and man, having enslaved the elements, remains himself a slave" (502–3). Shelley, of all the Romantic poets, was the one most interested in science, and what interests me in this passage with its talk of the poetic faculty (or imagination) is not how much Lopez is like a Romantic but how persistent (and trenchant) this critique of human dependence on rationality and human distrust of the imagination is. Karl Kroeber, in *Ecological Literary Criticism,* has a chapter on Shelley called "Shelley, the Socialization of Mind." According to Kroeber, we see in Shelley's poetry that the key to human interaction with nature is human interrelationships with other human beings. Nature and human consciousness cannot be separated. Lopez would agree, I think.

4. See Hitt. Hitt's "Toward an Ecological Sublime" is an interesting argument for the renewed usefulness of the romantic sublime. Hitt too makes an argument for wonder, contending that the human being faced with the utter inaccessibility of the natural world will wonder at its sublimity. I agree to an extent. But I think that we can take this a step further. To my mind the ecological sublime resides not only in the absolute otherness of the natural world but also in the human perception that we exist entirely in relation to this grand process beyond our final knowledge. So the wonder is not directed at the object but rather at our relation to it. And this relation is seen as an ongoing activity, changing and fluid.

5. See again Buell's discussion of "The Aesthetics of Relinquishment" (*Environmental Imagination* 143–79). Buell, like Lopez, believes that "environmental writing has to be able to imagine nonhuman agents as bona fide partners" (179).

6. Tallmadge remarks that Lopez considered entering a seminary or becoming a Trappist monk ("Barry Lopez" 549–50). See also Evans 67. Lopez decided that the life of a monk would be too easy. A strong commitment to the betterment of the world, especially the landscapes and peoples of North America, is certainly a spiritual mission for Lopez.

7. On reading and writing see again Poirier's *The Renewal of Literature:* "The experiences both of reading and of the writing it creates are more real, more present to consciousness, than are any prior circumstances that might have given rise to them. Our reading-writing brings into existence a moment in which we are actively there, but it is also a moment in which self-present identity is reportedly lost. For that reason we can say that man and not-man are in a simultaneous, occurrent state, that one cannot be severed from the other" (210–11). It seems as if the reading subject could theoretically participate with the environment in an ecology of reading-writing.

Conclusion

1. In this book, Buell complicates many of the issues he raised in *The Environmental Imagination,* including a long look at urban ecologies. Phillips, in *The*

Truth of Ecology, makes no mention of Buell's later work, although Buell endures much harsh criticism from Phillips.

2. In "Eco-logic: An Erotic of Nature," Glazebrook uses the term "eco-logic" in a slightly different way, although her critique of modern technological science and its objectification of the natural world ends up with strategies to help us behave toward the world in a more loving way, a strategy whose consequences Williams surely embraces. Glazebrook looks back to Aristotle and Plato for ways within the Western tradition to love nature. She argues that in an Aristotelian concept of nature as teleological, nature has final causes that are beyond our ken. Modern science, however, in its withdrawal from nature, understands final causes only as acts of "conscious intentions." Her argument concerning final causes is of great interest to me in the context of Dewey's work. I think that for Dewey the idea of interaction itself is a final cause—a final cause never finalized. For a pragmatist ecology, the very logic of loving participation in a world always in process could be the final, ecological word on experience. I see the ways that Glazebrook and I have used the idea of (eco)logic as complementary, and I am glad for the insights in her essay.

3. Roorda's proposal of a biopragmatist criticism asks what difference it would make in the lives of people and places to accept the terms of experience found in a book of nature writing. Roorda's pragmatist approach to reading and writing insists upon the reality of places and would pay close attention to consequences and to solving problems. In an environmental register, texts and readers engage in a kind of ecological inquiry that would be enhanced, as I hope becomes clear in the course of this section, by Dewey's work on logic.

4. This is, of course, the main theoretical contention in Holmes's biography of Muir.

5. Ross-Bryant sees Williams's project as a rewriting of natural history that includes the self. Farr sees Williams as a direct descendant of Edward Abbey in the tradition of "ecobiography." Allister points out that Williams's use of birds in *Refuge* illustrates the interrelation of human and animal that ultimately contributes strongly to her grief work in *Refuge*. It is not until Williams comes to understand the birds as the "relational other" necessary to the narration of her grief that the transformation of grief into hope can be accomplished.

6. In her essay on *Refuge* and ecofeminism, Glotfelty wishes finally that Williams would do more to bridge gender differences instead of exaggerating them. I see *Refuge* as less dualistic. Williams, indeed, radically shifts power to the feminine side of the spectrum, but it seems to me that she attempts, primarily through the thematization of family relations, to be more inclusive than Glotfelty suggests.

7. Gersdorf draws on two key texts: Audre Lorde's *Uses of the Erotic: The Erotic as Power* and Susan Griffin's *The Eros of Everyday Life*. Williams is consistently concerned with our erotic relationship to the land. In "Yellowstone: The Erotics of Place" she writes: "it is time for us to take off our masks, to step out from behind our personas—whatever they might be: educators, activists, biologists, geologists, writers, farmers, ranchers, and bureaucrats—and admit we are lovers, engaged in an erotics of place. Loving the land. Honoring its mysteries. Acknowledging, em-

bracing the spirit of place (there is nothing more legitimate and there is nothing more true" (*Unspoken Hunger* 84). As if to illustrate this truth, her next book, *Desert Quartet: An Erotic Landscape,* includes a passage where the female narrator has sexual relations with a river (22–24). Williams has consistently claimed a deeply felt relationship with the land, and she has also claimed that one of the most influential books on her life is Steinbeck's *To a God Unknown*. Interestingly, Joseph Wayne in *To a God Unknown* has a strong erotic attraction to the land, and Steinbeck's novel also includes a sex scene between the land and the main character (8). The principal difference, I think, is that in Williams we get the feeling that the land also enjoys itself, whereas in Steinbeck the relationship seems a bit more one-sided.

Works Cited

Abbey, Edward. *Desert Solitaire: A Season in the Wilderness*. 1968. New York: Simon & Schuster, 1990.
Abram, David. *The Spell of the Sensuous: Perception and Language in a More-Than-Human World*. New York: Vintage, 1996.
Alexander, Thomas M. "John Dewey and the Moral Imagination: Beyond Putnam and Rorty toward a Postmodern Ethics." *Transactions of the Charles S. Peirce Society* 29.3 (1993): 369–400.
———. *John Dewey's Theory of Art, Experience, and Nature: The Horizon of Feeling*. Albany: SUNY UP, 1987.
———. "Pragmatic Imagination." *Transactions of the Charles S. Peirce Society* 26.3 (1990): 325–49.
Allen, T. F. H. "Community Ecology." *Ecology*. Ed. Stanley I. Dodson, et al. New York: Oxford UP, 1998. 315–83.
Allister, Mark. *Refiguring the Map of Sorrow: Nature Writing and Autobiography*. Charlottesville: UP of Virginia, 2001.
Altieri, Charles. *Painterly Abstraction in Modernist American Poetry: The Contemporaneity of Modernism*. Cambridge: Cambridge: UP, 1989.
Anderson, Charles. *The Magic Circle of Walden*. New York: Holt, Rinehart, and Winston, 1968.
"Another Top 100 List: Now It's Nonfiction." *New York Times* 30 April 1999: E45.
Armbruster, Karla. "Rewriting a Genealogy with the Earth: Women and Nature in the Works of Terry Tempest Williams." *Southwestern American Literature* 21.1 (1995): 209–20.
Arnold, Jean. "'From So Simple a Beginning': Evolutionary Origins of U.S. Nature Writing." *ISLE* 10.1 (2003): 11–26.
Astro, Richard. *John Steinbeck and Edward Ricketts: The Shaping of a Novelist*. Minneapolis: U of Minnesota P, 1973.
Aton, James. "An Interview with Barry Lopez." *Western American Literature* 21.1 (1986): 3–17.

Attridge, Derek. "Innovation, Literature, Ethics: Relating to the Other." *PMLA* 114 (1999): 20–31.
Atwood, Margaret. *Strange Things: The Malevolent North in Canadian Literature*. Oxford: Clarendon, 1995.
Badè, William Frederic. *The Life and Letters of John Muir*. 2 vols. Boston: Houghton Mifflin, 1923–24.
Bakhtin, M. M. *The Dialogic Imagination*. Ed. Michael Holquist. Trans. Caryl Emerson and Michael Holquist. Austin: U of Texas P, 1981.
———. *Speech Genres and other Late Essays*. Ed. Vern W. McGee. Trans. Caryl Emerson and Michael Holquist. Austin: U of Texas P, 1986.
Beebe, William. Rev. of *Under the Sea Wind* by Rachel Carson. *Saturday Review of Literature* 27 December 1941: 5.
Beegel, Susan F., Susan Shillinglaw, and Wesley N. Tiffney Jr., eds. *Steinbeck and the Environment: Interdisciplinary Approaches*. Tuscaloosa: U of Alabama P, 1997.
Benson, Jackson J. *The True Adventures of John Steinbeck, Writer*. New York: Viking, 1984.
Berry, Wendell. "Speech after Long Silence." Bezner and Walzer 25–28.
Bezner, Kevin. "An Interview with John Haines." *Green Mountains Review* 6.2 (1993): 9–15.
———. "John Haines." *Updating the Literary West*. Fort Worth: Texas Christian UP, 1997. 274–77.
Bezner, Kevin, and Kevin Walzer, eds. *The Wilderness of Vision: On the Poetry of John Haines*. Brownsville, OR: Storyline, 1996.
Bhabha, Homi. *The Location of Culture*. London: Routledge, 1994.
"Board Says Mansion a Burden for YMCA." *Columbus Dispatch* 10 December 2000: D1.
Bone, Robert M. *The Geography of the Canadian North*. Toronto: Oxford UP, 1992.
Bonetti, Kay. "An Interview with Barry Lopez." *Missouri Review* 11.3 (1988): 59–77.
Boodin, John Elof. *A Realistic Universe*. New York: Macmillan, 1931.
Bortoft, Henri. *The Wholeness of Nature: Goethe's Way toward a Science of Conscious Participation in Nature*. Hudson, NY: Lindisfarne, 1996.
Boudreau, Gordon. *The Roots of Walden and the Tree of Life*. Nashville: Vanderbilt UP, 1990.
Branch, Michael P. "'Angel Guiding Gently': The Yosemite Meeting of Ralph Waldo Emerson and John Muir, 1871." *Western American Literature* 32.2 (1997): 126–49.
———. "John Muir's *My First Summer in the Sierra*." *ISLE* 11.1 (2004): 139–52.
———. "Telling Nature's Story: John Muir and the Decentering of the Romantic Self." Miller 99–122.
Brodwin, Stanley. "'The Poetry of Scientific Thinking': Steinbeck's *Log from the Sea of Cortez* and Scientific Travel Narrative." Beegel, Shillinglaw, and Tiffney 142–60.
Brooks, Paul. *The House of Life: Rachel Carson at Work*. Boston: Houghton Mifflin, 1972.

———. *Speaking for Nature: How Literary Naturalists from Henry Thoreau to Rachel Carson Have Shaped America*. San Francisco: Sierra Club Books, 1980.
Brown, Bill. *A Sense of Things: The Object Matter of American Literature*. Chicago: U of Chicago P, 2003.
Browne, Neil. "Activating the 'Art of Knowing': John Dewey, Pragmatist Ecology, and Environmental Writing." *ISLE* 11.2 (2004): 1–24.
Bryant, Lynn Ross. "The Self in Nature: Four American Autobiographies." *Soundings* 80.1 (1997): 83–103.
Buell, Lawrence. *The Environmental Imagination: Thoreau, Nature Writing, and the Formation of American Culture*. Cambridge: Harvard UP, 1995.
———. *The Future of Environmental Criticism*. Malden, MA: Blackwell, 2005.
———. *Writing for an Endangered World: Literature, Culture, and Environment in the U.S. and Beyond*. Cambridge: Harvard UP, 2001.
Calliou, Brian, and Cora Voyageur. "Aboriginal Economic Development and the Struggle for Self-Government." *Power and Resistance: Critical Thinking about Canadian Social Issues*. Ed. Wayne Antony and Les Samuelson. 2nd ed. Halifax: Fernwood, 1998. 115–34.
Carpenter, Frederick I. "The Philosophical Joads." *Steinbeck and His Critics: A Record of Twenty-five Years*. Ed. E. W. Tedlock and C. V. Wicker. Albuquerque: U of New Mexico P, 1957. 241–49.
Carson, Rachel. "A Battle in the Clouds." *St. Nicholas Magazine* September 1918: 1048.
———. *The Edge of the Sea*. 1955. Boston: Houghton Mifflin, 1998.
———. *Lost Woods: The Discovered Writing of Rachel Carson*. Ed. Linda Lear. Boston: Beacon, 1998.
———. "The Real World around Us." Carson, *Lost Woods* 147–63.
———. *The Sea around Us*. 1951. New York: Oxford, 1989.
———. *The Sense of Wonder*. New York: Harper and Row, 1965.
———. *Silent Spring*. Boston: Houghton Mifflin, 1962.
———. "Undersea." Carson, *Lost Woods* 3–11.
———. *Under the Sea Wind*. 1941. New York: Penguin, 1996.
Carson, Rachel, and Vanez T. Wilson. *Bear River: A National Wildlife Refuge. Conservation in Action #8*. Illustrated by Bob Hines. Washington, DC: U.S. Fish and Wildlife Service, Government Printing Office, 1950.
Casey, Edward S. "How to Get from Space to Place in a Fairly Short Stretch of Time: Phenomenological Prolegomena." *Senses of Place*. Ed. Steven Feld and Keith H. Basso. Santa Fe: School of American Research, 1996. 13–52.
Cassuto, David N. "Turning Wine into Water: Water as Privileged Signifier in *The Grapes of Wrath*." Beegel, Shillinglaw, and Tiffney 55–75.
Chaloupka, William. "John Dewey's Social Aesthetics as a Precedent for Environmental Thought." *Environmental Ethics* 9.3 (1987): 243–60.
Clarke, Bruce. "Science, Theory, and Systems: A Response to Glen A. Love and Jonathan Levin." *ISLE* 8.1 (2001): 149–65.
Clavigero, Don Francisco Javier. *The History of [Lower] California*. 1786. Trans. Sara

Lake. Ed. A. Gray. Bryn Mawr, CA, 1971. Reprinted from the Stanford University Press edition of 1937.

Coffin, Arthur. "An Interview with John Haines." *Jeffers Studies* 2.4 (1998): 47–56.

Cohen, Marty. Rev. of *The Stars, the Snow, the Fire* by John Haines. *Parnassus* 16.2 (1991): 143–59.

Cohen, Michael P. *The Pathless Way: John Muir and American Wilderness*. Madison: U of Wisconsin P, 1984.

Coleridge, Samuel Taylor. *Biographia Literaria*. 1815–17. *Samuel Taylor Coleridge: A Critical Edition of the Major Works*. Ed. H. J. Jackson. Oxford: Oxford UP, 1985. 155–482.

Coles, Romand. "Ecotones and Environmental Ethics: Adorno and Lopez." *In the Nature of Things: Language, Politics, and the Environment*. Ed. Jane Bennet and William Chaloupka. Minneapolis: U of Minnesota P, 1993. 226–49.

Collins, Francis, et al. "The Human Genome." *Nature* 15 February 2001: 813–958.

Cooperman, Matthew. "Traveling Papers: Assays in Poetry and Poetics." Diss. Ohio U, 1998.

——. "Wilderness and Witness: An Interview with John Haines." *QAE* 3 (1996): 111–27. Reprinted in Rogers 127–45.

Cronon, William. "Neither Barren Nor Remote." *New York Times* 28 February 2001: A25.

DeMott, Robert. *Steinbeck's Reading: A Catalogue of Books Owned and Borrowed*. New York: Garland, 1984.

——. *Steinbeck's Typewriter: Essays on His Art*. Troy, NY: Whitson, 1996.

Dewey, John. *John Dewey: The Later Works, 1925–1953*. Ed. Jo Ann Boydston. 17 vols. Carbondale: U of Southern Illinois P, 1981–90.

——. *John Dewey: The Middle Works, 1899–1924*. Ed. Jo Ann Boydston. 15 vols. Carbondale: U of Southern Illinois P, 1976–83.

Dimock, Wai Chee. "Literature for the Planet." *PMLA* 116 (2001): 173–88.

——. "A Theory of Resonance." *PMLA* 112 (1997): 1060–71.

Dodson, Stanley I., et al., eds. *Ecology*. New York: Oxford UP, 1998

Easterbrook, Donald J. *Surface Processes and Landforms*. Englewood Cliffs, NJ: Prentice Hall, 1993.

Elder, John, ed. *American Nature Writers*. 2 vols. New York: Scribner's, 1996.

——. *Imagining the Earth: Poetry and the Vision of Nature*. 2nd ed. Urbana: U of Illinois P, 1985.

Emerson, Ralph Waldo. *Nature*. 1836. *The Collected Works of Ralph Waldo Emerson*. Vol. 1. Ed. Alfred R. Ferguson. Cambridge: Harvard UP, 1971. 3–45.

——. "The Over-Soul." *Ralph Waldo Emerson: Essays and Poems*. London: Everyman, 1995. 130–45.

Englert, Peter. "Education of Environmental Scientists: Should We Listen to Steinbeck and Ricketts's Comments?" Beegel, Shillinglaw, and Tiffney 176–93.

Evans, Alice. "Leaning into the Light: An Interview with Barry Lopez." *Poets and Writers* 22.2 (1994): 62–79.

Evernden, Neil. *The Natural Alien: Humankind and Environment.* 1983. 2nd ed. Toronto: U of Toronto P, 1993.

Farr, Cecilia Konchar. "American Ecobiography." *Literature of Nature: An International Sourcebook.* Ed. Patrick Murphy. Chicago: Fitzroy Dearborn, 1998. 94–97.

Fensch, Thomas. *Steinbeck and Covici: The Story of a Friendship.* Middlebury, VT: Paul S. Erikkson, 1979.

Fesmire, Steven. *John Dewey and Moral Imagination: Pragmatism in Ethics.* Bloomington: Indiana UP, 2003.

Fleck, Richard F. *Henry Thoreau and John Muir among the Indians.* Hamden, CT: Archon Books, 1985.

Flores, Dan L. "Environmentalism and Multiculturalism." *Reopening the American West.* Ed. Hal K. Rothman. Tucson: U of Arizona P, 1998. 24–36.

Fox, Stephen. *John Muir and His Legacy: The American Conservation Movement.* Boston: Little, Brown, 1981.

Gartner, Carol B. *Rachel Carson.* New York: Frederick Ungar, 1983.

———. "When Science Becomes Literary Art." Waddell 103–25.

Gersdorf, Catrin. "Ecocritical Uses of the Erotic." *New Essays in Ecofeminist Literary Criticism.* Ed. Glynis Carr. Lewisburg, PA: Bucknell UP, 2000. 175–91.

Gifford, Terry, ed. *John Muir: His Life and Letters and Other Writings.* Seattle: Mountaineers, 1996.

———. "Muir's Ruskin: John Muir's Reservations about Ruskin Reviewed." Miller 123–50.

Glazebrook, Trish. "Eco-logic: An Erotic of Nature." *Rethinking Nature: Essays in Environmental Philosophy.* Ed. Bruce V. Foltz and Robert Frodeman. Bloomington: U of Indiana P, 2004. 95–113.

Glotfelty, Cheryll. "Cold War, *Silent Spring:* The Trope of War in Modern Environmentalism." Waddell 157–73.

———. "Flooding the Boundaries of Form: Terry Tempest Williams's Ecofeminist *Unnatural History.*" *Change in the American West: Exploring the Human Dimension.* Ed. Stephen Tchudi. Reno: U of Nevada P. 1996. 158–67.

———. "Rachel Carson." Elder, *Nature Writers* 1: 151–71.

Glotfelty, Cheryll, and Harold Fromm, eds. *The Ecocriticism Reader: Landmarks in Literary Ecology.* Athens: U of Georgia P, 1996.

Gouinlock, James. *John Dewey's Philosophy of Value.* New York: Humanities Press, 1972.

Gould, Stephen Jay. "Humbled by the Genome's Mysteries." *New York Times* 19 February 2001: A21.

Grace, Sherill. "Comparing Mythologies: Ideas of West and North." *Borderlands: Essays in Canadian-American Relations.* Ed. Robert Lecker. Toronto: ECW Press, 1991. 243–62.

Greenblatt, Stephen. "Resonance and Wonder." *Exhibiting Cultures: The Poetics and Politics of Museum Display.* Ed. Ivan Karp and Steven D. Lavine. Washington, DC: Smithsonian Institution P, 1991. 42–56.

Griffin, Susan. *The Eros of Everyday Life.* New York: Doubleday, 1995.

Gura, Philip F. *The Wisdom of Words: Language, Theology, and Literature in the New England Renaissance*. Middleton, CT: Wesleyan UP, 1981.

Hadley, Edith Jane. "John Muir's Views of Nature and Their Consequences." Diss. U of Wisconsin, 1956.

Haines, John. *Fables and Distances: New and Selected Essays*. Saint Paul: Graywolf, 1996.

———. *Living Off the Country: Essays on Poetry and Place*. Ann Arbor: U of Michigan P, 1981.

———. *New Poems*. Brownsville, OR: Storyline, 1990.

———. "The Rise of Nature Writing: America's Next Great Genre?" *Manoa* 4.2 (1992): 83–84.

———. *The Stars, the Snow, the Fire: Twenty-Five Years in the Alaska Wilderness*. Saint Paul: Graywolf, 1989. Reprinted 2000.

———. "The Writer as Alaskan." Lyon 366–80.

Hamill, Sam. "John Haines and the Place of Sense." Bezner and Walzer 171–81.

Hansen, Arlen J. "Right Men in the Right Places: The Meeting of Ralph Waldo Emerson and John Muir." *Western Humanities Review* 39.2 (1985): 165–72.

Harper, Kenn. "Hunters and High Finance." *Nunavut 99* <http://www.nunavut.com/nunavut99/english/ hunters.htm>.

Harper, Ralph. *Nostalgia: An Existential Exploration of Longing and Fulfillment in the Modern Age*. Cleveland: Western Reserve UP, 1966.

Hedgpeth, Joel W. *The Outer Shores, Part 1, Ed Ricketts and John Steinbeck Explore the Pacific Coast*. Eureka, CA: Mad River, 1978. 2 parts.

Hedin, Robert. "An Interview with John Haines." *Northwest Review* 27.2 (1989): 62–76.

Hefner, Philip. *The Human Factor: Evolution, Culture, and Religion*. Minneapolis: Fortress Press, 1993.

Hickman, Larry. "Nature as Culture: John Dewey's Pragmatic Naturalism." *Environmental Pragmatism*. Ed. Andrew Light and Eric Katz. London: Routledge, 1996. 50–72.

Hill, Jen. "Barry Lopez, *Arctic Dreams*." *Literature and the Environment*. Ed. George Hart and Scott Slovic. Westport, CT: Greenwood, 2004. 129–41.

Hitt, Christopher. "Toward an Ecological Sublime." *New Literary History* 30.3 (1999): 603–23.

Hoagland, Edward. *Notes from the Century Before: A Journal from British Columbia*. New York: Random House, 1969.

Holmes, Steven J. *The Young John Muir: An Environmental Biography*. Madison: U of Wisconsin P, 1999.

Hönnighausen, Lothar. *Grundprobleme der Englischen Literaturtheorie des Neunzehnten Jahrhunderts*. Darmstadt: Wissenschaftliche Buchgesellschaft, 1977.

Hook, Sidney. Introduction. Dewey, *Middle Works* 8: ix–xxxvi.

James, William. *Essays in Radical Empiricism*. New York: Longman, Green, 1912.

———. *Pragmatism*. 1907. Cambridge: Harvard UP, 1975.

Jensen, Derrick. *Listening to the Land: Conversations about Nature, Culture, and Eros.* San Francisco: Sierra Club, 1995.

Kestenbaum, Victor. *The Grace and the Severity of the Ideal: John Dewey and the Transcendent.* Chicago: U of Chicago P, 2002.

———. *The Phenomenological Sense of John Dewey: Habit and Meaning.* Atlantic Highlands, NJ: Humanities Press, 1977.

Kimes, William F., and Maymie Kimes. *John Muir: A Reading Bibliography.* Fresno: Panorama Books, 1986.

Kircher, Cassandra. "Rethinking Dichotomies in Terry Tempest Williams's *Refuge.*" *ISLE* 3.1 (1996): 97–114.

Kollin, Susan. "The Wild, Wild North: Nature Writing, Nationalist Ecologies, and Alaska." *American Literary History* 12.1–2 (2000): 41–78.

Kroeber, Karl. *Ecological Literary Criticism: Romantic Imagining and the Biology of Mind.* New York: Columbia UP, 1994.

———. "Ecology and American Literature: Thoreau and Un-Thoreau." *American Literary History* 9.2 (1997): 309–28.

Kuhn, Thomas. *The Structure of Scientific Revolutions.* Chicago: U of Chicago P, 1962.

Lear, Linda. *Rachel Carson: Witness for Nature.* New York: Henry Holt, 1997.

Leopold, Aldo. *A Sand County Almanac and Sketches Here and There.* Oxford: Oxford UP, 1949.

Levin, Jonathan. "Between Science and Anti-Science: A Response to Glen A. Love." *ISLE* 7.1 (2000): 1–7.

Lopez, Barry. *About This Life: Journeys on the Threshold of Memory.* New York: Knopf, 1998.

———. *Arctic Dreams: Imagination and Desire in a Northern Landscape.* New York: Scribner's, 1986.

———. *Crossing Open Ground.* New York: Vintage, 1989.

———. "Learning to See." *DoubleTake* 4.2 (1998): 73–79.

———. *The Rediscovery of North America.* New York: Vintage, 1990.

———. "We Are Shaped by the Sound of Wind, the Slant of Sunlight." *High Country News* 14 September 1998: 1, 10–11.

Lorde, Audre. *Uses of the Erotic: The Erotic as Power.* Trumansburg, NY: Out & Out Books, 1978.

Love, Glen. *Practical Ecocriticism: Literature, Biology, and the Environment.* Charlottesville: U of Virginia P, 2003.

———. "Science, Anti-Science, and Ecocriticism." *ISLE* 6.1 (1999): 65–81.

Lueders, Edward, ed. *Writing Natural History: Dialogues with Authors: Barry Lopez and Edward O. Wilson, Robert Finch and Terry Tempest Williams, Gary Paul Nabhan and Ann Zwinger, Paul Brooks and Edward Lueders.* Salt Lake City: U of Utah P, 1989.

Lyman, John. Rev. of *Sea of Cortez* by John Steinbeck and Edward F. Ricketts. *American Neptune* 2.2 (1942): 183.

Lyon, Thomas J., ed. *This Incomperable Lande: A Book of American Nature Writing.* New York: Penguin, 1989.

Matthews, Freya. "Ecofeminism and Deep Ecology." Merchant 235–45.

Mazel, David. *American Literary Environmentalism.* Athens: U of Georgia P, 2000.

McCay, Mary A. *Rachel Carson.* New York: Twayne, 1993.

McDonald, Hugh P. *John Dewey and Environmental Philosophy.* Albany: SUNY P, 2004.

McFague, Sallie. *The Body of God: An Ecological Theology.* Minneapolis: Fortress Press, 1993.

McPhee, John. *Coming into the Country.* New York: Farrar, Straus and Giroux, 1976.

Merchant, Carolyn, ed. *Key Concepts in Critical Theory: Ecology.* Atlantic Highlands, NJ: Humanities Press, 1994.

Merleau-Ponty, Maurice. *Phenomenology of Perception.* Trans. Colin Smith. London: Routledge, 1962.

Miller, Sally M., ed. *John Muir in Historical Perspective.* New York: Peter Lang, 1999.

Muir, John. "For the Boston Recorder. The Calypso Borealis. Botanical Enthusiasm. From Prof. J. D. Butler." *Boston Recorder* 21 December 1866: 1.

———. *The Mountains of California.* 1894. Garden City,: Doubleday, 1961.

———. *My First Summer in the Sierra.* 1911. New York: Penguin, 1997.

———. *Our National Parks.* Boston: Houghton Mifflin, 1901.

———. "Out of the Wilderness." *Atlantic Monthly* February 1913: 266–77.

———. *The Story of My Boyhood and Youth.* Boston: Houghton Mifflin, 1913.

———. *A Thousand-Mile Walk to the Gulf.* Ed. William Frederic Badè. Boston: Houghton Mifflin, 1917.

———. *Travels in Alaska.* 1915. San Francisco: Sierra Club, 1988.

———. "Twenty Hill Hollow." *Overland Monthly* July 1872: 80–86.

Nash, Roderick. *The Rights of Nature: A History of Environmental Ethics.* Madison: U of Wisconsin P, 1989

———. *Wilderness and the American Mind.* 4th ed. New Haven: Yale UP, 2001.

Norwood, Vera. "Heroines of Nature: Four Women Respond to the American Landscape." Glotfelty and Fromm 323–50.

———. *Made from This Earth: American Women and Nature.* Chapel Hill: U of North Carolina P, 1993.

———. "The Nature of Knowing: Rachel Carson and the American Environment." *Signs* 12.4 (1987): 740–60.

Novak, Barbara. *Nature and Culture: American Landscape Painting 1825–1875.* Rev. ed. New York: Oxford UP, 1995.

Nowak, Elke. "Inuit—Die Menschen der kanadischen Arktis." *Zeitschrift für Kanada-Studien* 34.2 (1998): 5–24.

Oates, Joyce Carol. "Against Nature." *On Nature: Nature, Landscape, and Natural History.* Ed. Daniel Halpern. San Francisco: Northpoint, 1986. 236–43.

O'Connell, Nicholas. "At One with the Natural World: Barry Lopez's Adventure with the Word and the Wild." *Commonweal* 24 March 2000: 11–17.

Oelschlaeger, Max. *Caring for Creation: An Ecumenical Approach to the Environmental Crisis*. New Haven: Yale UP, 1994.

O'Grady, John. *Pilgrims to the Wild: Everett Ruess, Henry David Thoreau, John Muir, Clarence King, Mary Austin*. Salt Lake City: U of Utah P, 1993.

Orr, David. *Ecological Literacy: Education and the Transition to a Postmodern World*. Albany: SUNY P, 1992.

Papin, Liliane. "This Is Not a Universe: Metaphor, Language, and Representation." *PMLA* 107 (1992): 1253–65.

Paul, Sherman. *For Love of the World: Essays on Nature Writers*. Iowa City: U of Iowa P, 1992.

———. *The Shores of America: Thoreau's Inward Exploration*. Urbana: U of Illinois P, 1958.

Peattie, Donald Culross. Rev. of *Sea of Cortez* by John Steinbeck and Edward F. Ricketts. *Saturday Review of Literature* 27 December 1941: 5–7.

Peck, H. Daniel. *Thoreau's Morning Work: Memory and Perception in* A Week on the Concord and Merrimack Rivers, *the Journal, and* Walden. New Haven: Yale UP, 1990.

Pelly, David F. "Dawn of Nunavut." *Canadian Geographic* March/April 1993: 20–29.

Perez, Betty L. "The Form of the Narrative Section of *Sea of Cortez*: A Specimen Collected from Reality." *Steinbeck's Travel Literature: Essays in Criticism*. Ed. Tetsumaro Hayashi. Steinbeck Monograph Series no. 10. Muncie, IN: Steinbeck Society of America, 1980. 47–55.

Peterson, Anna L. *Being Human: Ethics, Environment, and Our Place in the World*. Berkeley: U of California P, 2001.

Philippon, Daniel J. *Conserving Words: How American Nature Writers Shaped the Environmental Movement*. Athens: U of Georgia P, 2004.

Phillips, Dana. *The Truth of Ecology: Nature, Culture, and Literature in America*. Oxford: Oxford UP, 2003.

Pinchot, Gifford. *Breaking New Ground*. New York: Harcourt, Brace, 1947.

Plumwood, Val. "Ecosocial Feminism as a General Theory of Oppression." Merchant 207–19.

Poirier, Richard. *The Renewal of Literature: Emersonian Reflections*. New York: Random House, 1987.

Railsback, Brian E. *Parallel Expeditions: Charles Darwin and the Art of John Steinbeck*. Moscow: U of Idaho P, 1995.

Richardson, Robert D. *Henry Thoreau: A Life of the Mind*. Berkeley: U of California P, 1986.

Ricketts, Edward F., and Jack Calvin. *Between Pacific Tides*. 1939. Revised by Joel W. Hedgpeth and D. W. Phillips. Stanford: Stanford UP, 1985.

Robinson, David. *Natural Life: Thoreau's Worldly Transcendentalism*. Ithaca: Cornell UP, 2004.

Rogers, Steven B., ed. *A Gradual Twilight: An Appreciation of John Haines*. Fort Lee, NJ: CavanKerry Press, 2003.

Roorda, Randall. *Dramas of Solitude: Narratives of Retreat in American Nature Writing*. Albany: SUNY P, 1998.

Rosenthal, Sandra B., and Patrick L. Bourgeois. *Pragmatism and Phenomenology: A Philosophic Encounter*. Amsterdam: B.R. Grüner, 1980.

Ross-Bryant, Lynn. "The Self in Nature: Four American Autobiographies." *Soundings* 80.1 (1997): 83–104.

Rueckert, William. "Barry Lopez and the Search for a Dignified and Honorable Relationship with Nature." *Earthly Words: Essays on Contemporary American Nature and Environmental Writers*. Ed. John Cooley. Ann Arbor: U of Michigan P, 1994. 137–64.

Runte, Alfred. *National Parks: The American Experience*. 2nd ed. Lincoln: U of Nebraska P, 1987.

Scheese, Don. *Nature Writing: The Pastoral Impulse in America*. New York: Twayne, 1996.

Shelley, Percy Bysshe. *A Defence of Poetry. Shelley's Poetry and Prose*. Ed. Donald H. Reiman and Sharon B. Powers. New York: Norton, 1977. 480–508.

Shusterman, Richard. *Pragmatist Aesthetics: Living Beauty, Rethinking Art*. 2nd ed. Lanham, MD: Rowman and Littlefield, 2000.

Siporin, Ona. "Terry Tempest Williams and Ona Siporin: A Conversation." *Western American Literature* 31.2 (1996): 99–113.

Slovic, Scott. *Seeking Awareness in American Nature Writing: Henry Thoreau, Annie Dillard, Edward Abbey, Wendell Berry, Barry Lopez*. Salt Lake City: U of Utah P, 1992.

Snyder, Gary. "Ecology, Literature, and the New World Disorder." *ISLE* 11.1 (2004): 1–13.

———. *The Practice of the Wild*. San Francisco: North Point, 1990.

Solnit, Rebecca. *Savage Dreams: A Journey into the Hidden Wars of the American West*. San Francisco: Sierra Club, 1994.

Sowell, Thomas. "*Silent Spring* Sparked Lots to Keep Quiet." *Columbus Dispatch* 9 June 2001: A12.

Stanley, Millie. *The Heart of John Muir's World: Wisconsin, Family, and Wilderness Discovery*. Madison: Prairie Oak, 1995.

St. Armand, Barton Levi. "The Book of Nature and American Nature Writing: Codex, Index, Contexts, Prospects." *ISLE* 4.1 (1997): 29–42.

Stein, Roger B. *John Ruskin and Aesthetic Thought in America*. Cambridge: Harvard UP, 1967.

Steinbeck, John. *East of Eden*. New York: Viking, 1952.

———. *The Grapes of Wrath*. 1939. New York: Penguin, 1992.

———. *Steinbeck: A Life in Letters*. Ed. Elaine Steinbeck and Robert Wallstein. New York: Viking, 1975.

———. *To a God Unknown*. 1933. New York: Penguin, 1995.

Steinbeck, John, and Edward F. Ricketts. *Sea of Cortez: A Leisurely Journal of Travel and Research*. New York: Viking, 1941. Reprint, Mamaroneck, NY: Paul P. Appel, 1971.

Stewart, Frank. *A Natural History of Nature Writing*. Washington, DC: Island Press, 1995.

Sweet, Timothy. *American Georgics: Economy and Environment in Early American Literature*. Philadelphia: U of Pennsylvania P, 2002.

Tallmadge, John. "Barry Lopez." Elder, *Nature Writers* 1: 549–68.

———. "Beyond the Excursion: Initiatory Themes in Annie Dillard and Terry Tempest Williams." *Reading the Earth: New Directions in the Study of Literature and Environment*. Ed. Michael P. Branch et al. Moscow: U of Idaho P, 1998. 197–207.

Tanner, Tony. *The Reign of Wonder: Naivety and Reality in American Literature*. Cambridge: Cambridge UP, 1965.

Tatum, Stephen. "Topographies of Transition in Western American Literature." *Western American Literature* 32.4 (1998): 311–52.

Taylor, Bob Pepperman. "John Dewey and Environmental Thought: A Response to Chaloupka." *Environmental Ethics* 12 (1990): 175–84.

Thoreau, Henry David. *The Journal of Henry David Thoreau*. 6 October 1857. Ed. Bradford Torrey and Francis H. Allen. 14 vols. bound as 2. Boston: Houghton Mifflin, 1906. 2: 1199–1200.

———. "Natural History of Massachusetts." 1842. *The Writings of Henry David Thoreau*. Ed. Bradford Torrey. Vol. 5. New York: Houghton Mifflin, 1906. 103–31.

———. *Walden and Civil Disobedience*. 2nd ed. Ed. William Rossi. New York: Norton, 1992.

———. "Walking." *The Writings of Henry David Thoreau*. Ed. Bradford Torrey. Vol. 5. Boston: Houghton Mifflin, 1906. 205–48.

Timmerman, John H. "Steinbeck's Environmental Ethic: Humanity in Harmony with the Land." Beegel, Shillinglaw, and Tiffney 310–22.

Trueblood, Valerie. "One to Whom the Great Announcements Are Made." *American Poetry Review* 32.1 (2003): 47–50.

Turner, Frederick. *Rediscovering America: John Muir in His Time and Ours*. San Francisco: Sierra Club, 1985.

Turner, Jack. *The Abstract Wild*. Tucson: U of Arizona P, 1996.

Union of Concerned Scientists. "Invasive Species Degrading Alaska's Ecosystems" <http://www.ucsusa.org/news/press_release/new-report-finds-invasive-species-degrading-alaskas-ecosystems.html>.

United States Hydrographic Office. H.O. Publication #84, *Coast Pilot: Sailing Directions for the West Coasts of Mexico and Central America, 1937*.

Usher, Peter. "The North: One Land, Two Ways of Life." *Heartland, Hinterland: A Regional Geography of Canada*. Ed. Larry McCann and Angus Gunn. 3rd ed. Scarborough, Ontario: Prentice Hall Canada, 1998. 357–94.

Van Wyck, Peter C. *Primitives in the Wilderness: Deep Ecology and the Missing Human Subject*. Albany: SUNY P, 1997.

Ventor, J. Craig, et al. "The Sequence of the Human Genome." *Science* 16 February 2001: 1304–51.

Vlessides, Mike. "A Public Government." *Nunavut 99* <http://www.nunavut.com/ nunavut99/english/ public_gov.htm>.

Waddell, Craig, ed. *And No Birds Sing: Rhetorical Analyses of Rachel Carson's Silent Spring.* Carbondale: U of Southern Illinois P, 2000.

Wallace, David Raines. "What Is to Be Done with the Biosphere?" *Manoa* 4.2 (1992): 94–95.

Walls, Laura Dassow. *Seeing New Worlds: Henry David Thoreau and Nineteenth-Century Natural Science.* Madison: U of Wisconsin P, 1995.

Wesling, Donald. "The Poetics of Description: John Muir and Ruskinian Descriptive Prose." *Prose Studies 1800–1900* 1.1 (1977): 37–44.

West, Michael. *Transcendental Wordplay: America's Romantic Punsters and the Search for the Language of Nature.* Athens: Ohio UP, 2000.

Westbrook, Robert B. *Democratic Hope: Pragmatism and the Politics of Truth.* Ithaca: Cornell UP, 2005.

———. *John Dewey and American Democracy.* Ithaca: Cornell UP, 1991.

Weston, Anthony. "Beyond Intrinsic Value: Pragmatism in Environmental Ethics." *Environmental Ethics* 7.3 (1985): 321–39.

White, Fred D. "Rachel Carson: Encounters with the Primal Mother." *North Dakota Quarterly* 59.2 (1991): 184–97.

White, Lynn, Jr. "The Historical Roots of Our Ecological Crisis." *Science* 10 March 1967: 1203–7. Reprinted in Glotfelty and Fromm 3–14.

White, Richard. "'Are You an Environmentalist or Do You Work for a Living?' Work and Nature." *Uncommon Ground: Toward Reinventing Nature.* Ed. William Cronon. New York: Norton, 1995. 171–85.

Wild, Peter. *John Haines.* Boise State University Western Writers Series 68. Boise: Boise State, 1985.

———. Rev. of *The Stars, the Snow, the Fire* by John Haines. *Sierra* 75.2 (1990): 92–94.

Wilkins, Thurman. *John Muir: Apostle of Nature.* Oklahoma Western Biographies vol. 8. Norman: U of Oklahoma P, 1995.

Williams, Raymond. *The Country and the City.* New York: Oxford UP, 1973.

Williams, Terry Tempest. *Desert Quartet: An Erotic Landscape.* Illustrations by Mary Frank. New York: Pantheon, 1995.

———. "A Eulogy for Edward Abbey." *Resist Much Obey Little: Remembering Ed Abbey.* Ed. James R. Hepworth and Gregory McNamee. San Francisco: Sierra Club, 1996. 199–203.

———. *The Open Space of Democracy.* Illustrations by Mary Frank. Great Barrington, MA: Orion Society, 2004.

———. "The Outside Canon." *Outside* 21.5 (1996): 71.

———. *Refuge: An Unnatural History of Family and Place.* New York: Vintage, 1991.

———. "The Spirit of Rachel Carson." *Audubon* 94.4 (1992): 104–7.

———. *An Unspoken Hunger.* New York: Pantheon, 1994.

Wilson, Edmund. *The Boys in the Back Room: Notes on California Novelists.* San Francisco: Colt, 1941.

Wolfe, Linnie Marsh, ed. *John of the Mountains: The Unpublished Journals of John Muir.* Boston: Houghton Mifflin, 1938.

———. *Son of the Wilderness: The Life of John Muir.* New York: Knopf, 1946. Reprint, Madison: U of Wisconsin P, 1978.

Worster, Donald. *Nature's Economy: A History of Ecological Ideas.* Cambridge: Cambridge UP, 1977.

Wu, J., and O. L. Loucks. "From Balance of Nature to Hierarchical Patch Dynamics: A Paradigm Shift in Ecology." *Quarterly Review of Biology* 70 (1995): 439–66.

Wynn, Graeme. *Remaking the Land God Gave to Cain: A Brief Environmental History of Canada.* Canada House Lecture Series no. 62. Toronto: Canada House, 1999.

Zuelke, Karl. "The Ecopolitical Space of Refuge." *Surveying the Literary Landscapes of Terry Tempest Williams.* Ed. Katherine R. Chandler and Melissa A. Goldthwaite. Salt Lake City: U of Utah P, 2003. 239–50.

Index

Abbey, Edward, 128, 183, 203n5
Abram, David, 65, 194n14
Agassiz, Louis: *Études sur les glaciers*, 28, 190n7
Alexander, Thomas M.: "John Dewey and the Moral Imagination: Beyond Putnam and Rorty toward a Postmodern Ethics," 2, 150, 187n1; *John Dewey's Theory of Art, Experience, and Nature: The Horizon of Feeling*, 12–13, 71, 125; "Pragmatic Imagination," 158
Allee, W. C., 53
Allen, T. F. H., 8, 156, 164
Allister, Mark, 92, 187n3, 203n5; *Refiguring the Map of Sorrow: Nature Writing and Autobiography*, 2
Altieri, Charles, 200n4
Anderson, Charles, 189n9
anthropocentrism, 29
Arctic Wildlife Refuge, 134
Armbruster, Karla, 175
Arnold, Jean, 53; "'From So Simple a Beginning': Evolutionary Origins of U.S. Nature Writing," 187n4
A Room of One's Own (Virginia Woolf), 80
art of knowing, 3–7, 19, 27, 34, 38, 39, 52, 54, 55, 63, 74, 84, 94, 95, 126, 155, 165, 184, 185
Astro, Richard, 53, 60

Atlantic Monthly, 82
Aton, James, 157, 162
Attridge, Derek, 11–12, 16, 115, 132, 149, 176, 184
Atwood, Margaret, 168
Audubon, John James, 99

Badè, William Frederic, 36–37, 190n1
Bakhtin, M. M., 75
Beebe, William, 79, 193n11
Benson, Jackson J., 53, 192n5
Berry, Wendell, 119
Bezner, Kevin, 114, 128, 139
Bhabha, Homi, 75
Bloom, Harold, 55
Bly, Robert, 114
Bonetti, Kay, 157
Boodin, John Elof, 53, 61
Bortoft, Henri, 35
Boudreau, Gordon, 189n9
Bourgeois, Patrick L., 194n14
Branch, Michael P., 63, 190n2, 191n15
Bridger, Jim, 178
Broch, Hermann, 112
Brodwin, Stanley, 60, 193n7
Brooks, Paul, 199n29; *The House of Life: Rachel Carson at Work*, 98; *Speaking for Nature: How Literary Naturalists from Henry Thoreau to Rachel Carson Have Shaped America*, 106

Brown, Bill, 100
Browne, Neil, 187n4
Buell, Lawrence, 191n10, 202n5; *The Environmental Imagination: Thoreau, Nature Writing, and the Formation of American Culture*, 16, 18, 28–29, 71, 103, 144, 146, 198n27; *The Future of Environmental Criticism*, 171; *Writing for an Endangered World: Literature, Culture, and Environment in the U.S. and Beyond*, 90, 171, 202n1
Butler, James Davie, 48

Calliou, Brian, 167
Calvin, Jack, 197n16
Carpenter, Frederic: "The Philosophical Joads," 51
Carr, Jeanne, 48
Carson, Rachel, 7, 9, 58, 79–110, 112, 136, 145, 172, 183, 193n11; "A Battle in the Clouds," 81; *Bear River: A National Wildlife Refuge*, 178–79; *The Edge of the Sea*, 79, 84–85, 92, 96–108; "The Real World Around Us," 196n8; *The Sea Around Us*, 79, 87–96, 99, 183; *The Sense of Wonder*, 196n10; *Silent Spring*, 79, 80, 81, 107–10, 195n3, 197n23 198n28; "Undersea," 82–83, 105; *Under the Sea Wind*, 79, 82, 83–88, 92
Carter, Jimmy, 199n30
Casey, Edward S., 171, 174
Cassuto, David N., 192n3
Chaloupka, William, 75, 97, 98, 107
Clarke, Bruce, 187n4
Clavigero, Don Francisco Javier: *The History of Lower California*, 72–73
Coast Pilot, 70, 71, 72–73, 75, 88–89, 183
Coffin, Arthur, 132
Cohen, Marty, 114
Cohen, Michael P., 22, 25, 31, 190n1, 190n5
Coleridge, Samuel Taylor, 91, 191n18, 197n18
Coles, Romand, 146, 159, 165

Collins, Francis, 193n9
community, 2, 6, 19, 26–27, 36, 48, 64–67, 71, 75, 77, 79, 100, 107, 109, 116, 117, 122, 125, 129, 139–41, 158, 160, 164, 170, 172, 181, 184–85
conservation and preservation, 27
Cooperman, Matthew, 136, 142, 145
Covici, Pat, 59
Cronon, William, 134, 179

Darwin, Charles, 28, 56; *On the Origin of Species*, 38; *The Voyage of the Beagle*, 54. See also Dewey, John, "The Influence of Darwinism on Philosophy"
Delany, Patrick, 22
democracy, 2, 3, 5–6, 7, 9, 11, 18, 19, 35, 36, 48, 51, 59, 65, 83, 94–95, 103–4, 108, 113, 114, 116, 125, 140–41, 152, 158, 164–66, 172, 176, 177–78, 184–85, 197n4
DeMott, Robert: *Steinbeck's Reading: A Catalogue of Books Owned and Borrowed*, 52, 77, 192n4; *Steinbeck's Typewriter: Essays on His Art*, 192n1
Dewey, John, 1–19, 57, 61, 74, 79, 84, 90, 92, 93, 96, 100, 101, 114, 137, 140, 148, 163, 164, 165, 168, 175, 188n4, 190n8, 197n24; *A Common Faith*, 52, 192n4; "Americanism and Localism," 179; *Art as Experience*, 10–13, 15–16, 18, 19, 23, 29, 33, 34–35, 41, 43, 44, 52, 61, 64, 66–67, 69–70, 71, 77, 87, 91, 92, 93–94, 99, 100, 101, 104, 113, 114, 116, 120, 132, 145, 146, 151, 163, 165, 171, 172, 185; "Creative Democracy—The Task Before Us," 94–95; *Democracy and Education*, 3, 4, 9; "Democracy and Educational Administration," 189n13; "The Development of American Pragmatism," 51–52, 124; *Experience and Nature*, 4–5, 9–13, 16–17, 23, 31, 39, 43, 49, 52, 61, 63, 64, 106, 115, 122, 127, 129, 145, 157–58, 189n11;

"Freedom," 65; "The Influence of Darwinism on Philosophy," 28, 38, 53–54, 99; *Knowing and the Known*, 34; *Logic: The Theory of Inquiry*, 55–56, 172–74, 176–77, 179, 180–81, 183, 184–85; "Philosophy and Democracy," 177; "Philosophy's Future in Our Scientific Age," 4; *The Public and Its Problems*, 5, 65, 95, 130, 176, 179, 184; *The Quest for Certainty*, 6, 63, 156, 164

Dimock, Wai Chee, 176, 177; "A Theory of Resonance," 175–76; "Literature of the Planet," 175–76

Dodson, Stanley I., 188n8

ecocriticism, 7–8, 17, 188n7
ecological sublime, 151–52, 202n4. *See also* Hitt, Christopher
ecotone, 3, 9, 14, 18, 27, 34, 42, 47, 59, 71, 73, 75, 86, 87, 91, 98, 100, 101, 102, 104, 117, 120, 129, 145–46, 152, 154, 155, 160, 163, 165–66, 167, 182, 184–85, 187n3
Education of Henry Adams, The (Henry Adams), 80
Ehrlich, Gretel, 43
Elder, John, 155
Emerson, Ralph Waldo, 36–39, 40, 43, 51, 68, 191n15
Englert, Peter, 74
ethics, 9, 12, 17, 42, 54, 55, 68, 70, 71, 79, 80, 91, 92, 93, 95, 96, 97, 102, 103–4, 106, 108–110, 115, 132–33, 136, 152, 153, 159–64, 167
Evans, Alice, 150, 202n6
Evernden, Neil, 33, 191n10
experience, 2, 9–13, 15–16, 19, 23–24, 34–35, 49, 51, 55, 63, 65, 69, 71, 74, 75, 77, 79, 81, 83, 86, 93, 94–95, 97–98, 100, 101, 104, 105, 107, 113, 115, 119, 120–21, 145, 147, 152, 154, 170, 172–74, 177, 184, 190n8

Farr, Cecilia Konchar, 203n5
Fensch, Thomas, 71

Fesmire, Steven, 187n1
Fiedler, Leslie, 55
Fleck, Richard F., 36, 191n12
Flores, Dan L.: "Environmentalism and Multiculturalism," 18
Fox, Stephen, 37, 38, 190n1
Freeman, Dorothy, 107

Gartner, Carol B., 195n1, 196n9, 196n14
Gersdorf, Catrin, 182, 203n7
Gifford, Terry, 37; "Muir's Ruskin: John Muir's Reservations about Ruskin Reviewed," 191n17
Glazebrook, Trish: "Eco-logic: An Erotic of Nature," 203n2
Glotfelty, Cheryll, 96, 196n15, 203n6
Goethe, Johann Wolfgang von, 189n10
Gouinlock, James, 156
Gould, Stephen Jay, 56–57
Grace, Sherill: "Comparing Mythologies: Ideas of West and North," 153, 155
Greenblatt, Stephen, 184
Griffin, Susan: *The Eros of Everyday Life*, 203n7
Gruening, Ernest, 195n3
Gura, Philip F., 189n9

Hadley, Edith Jane, 191n17
Haeckel, Ernst, 22
Haines, John, 7, 26, 112–42, 145, 148, 170, 183; *New Poems*, 133, 141; "On a Certain Attention to the World," 122, 127; *The Stars, the Snow, the Fire*, 112–42, 145; *Winter News*, 112, 144; "The Writer as Alaskan," 117
Hamelin, Louis-Edmond, 149
Hamill, Sam, 139
Hansen, Arlen J., 191n15
Harper, Ken, 166
Harper, Ralph: *Nostalgia*, 126–28, 132
Hedgpeth, Joel, 54
Hedin, Robert, 118, 119
Hetch Hetchy Valley, 24–25
Hickman, Larry, 17, 189n11

Higgins, Elmer, 82
Hill, Jen, 160, 161
Hitt, Christopher, 202n4. *See also* ecological sublime
Hoagland, Edward: *Notes from the Century Before: A Journal from British Columbia*, 201n7
Holmes, Steven J., 22–23, 24, 25, 26, 48, 190n1
Hönnighausen, Lothar, 190n6, 191n16
Hook, Sidney, 172
Hopkins Marine Station, 53
Humphrey, Hubert, 195n3

instrumentalism, 51–52, 146–47

James, William, 10, 12, 23, 51, 61, 77, 189n11; *Pragmatism*, 52; *The Principles of Psychology*, 52; *The Varieties of Religious Experience*, 80, 192n4
Jena, Germany, 22
Jensen, Derrick, 183
Jewett, Sarah Orne: *The Country of the Pointed Firs*, 100
John Burroughs Award (Carson), 90
Johns Hopkins University, 81, 82
Johnson, Robert Underwood, 25

Kestenbaum, Victor, 84, 85, 194n14
Kimes, Maymie, 22, 48
Kimes, William F., 22, 48
Kircher, Cassandra, 179–80
Kollin, Susan, 153, 163
Kroeber, Karl, 150, 202n3
Kuhn, Thomas, 195n4

Lear, Linda, 85, 91, 93, 96, 107, 108, 109, 110, 195n3
Leopold, Aldo, 79, 96, 160
Levin, Jonathan, 187n4
Lopez, Barry, 7, 144–68, 183; *About This Life: Journeys on the Threshold of Memory*, 67–68; *Arctic Dreams: Imagination and Desire in a Northern Landscape*, 49, 67, 144–68, 201n7; *Crossing Open Ground*, 145, 197n20; "Learning to See," 159–60; *Of Wolves and Men*, 144; *The Rediscovery of North America*, 136, 161, 201n1; "We Are Shaped by the Sound of Wind, the Slant of Sunlight," 164
Lorde, Audre, 48; *The Uses of the Erotic: The Erotic as Power*, 203n7
Lorenz, Edward, 64
Loucks, O. L., 188n8; "From Balance of Nature to Hierarchical Patch Dynamics: A Paradigm shift in Ecology," 152–53. *See also* metastability; patch dynamics; Wu, J.
Love, Glen, 187n4; *Practical Ecocriticism: Literature, Biology, and the Environment*, 187n4
Lueders, Edward, 162
Lyell, Charles, 9–10; *Principles of Geology*, 38
Lyman, John, 70, 198n26
Lyon, Thomas J., 80, 106

McCay, Mary A., 198n28
McDonald, Hugh P., 52, 147, 198n25
McPhee, John: *Coming into the Country*, 201n7
Matthews, Freya, 73
Mazel, David, 17
Merleau-Ponty, Maurice, 194n14
metastability, 13, 152–53, 161. *See also* Loucks, O. L.; patch dynamics; Wu, J.
Monsanto, 110
Mt. Ritter, 31–32, 34–35
Muir, John, 7, 17, 22–49, 55, 62, 73, 79, 81, 94, 112, 183, 191n11, 191n13, 191n14, 191n15; "For the Boston Recorder: THE CALYPSO BOREALIS. Botanical Enthusiasm. From Prof. J. D. Butler," 48; *The Mountains of California*, 29–32; *My First Summer in the Sierra*, 22–49, 51; *Our National Parks*, 37; *The Story of My Boyhood and Youth*, 190n1; *A Thousand-Mile Walk to the Gulf*, 30, 36, 43; "Twenty Hill Hollow," 36–37

Nash, Roderick, 190n4; *Wilderness and the American Mind*, 25
National Book Award: Carson, 90; Lopez, 147
Nature, 56
nature as book, 17, 28, 190n7
New York Times, 134; "Another Top 100," 80
Norwood, Vera, 89, 95; *Made from This Earth: American Women and Nature*, 80
nostalgia, 126–29. *See also* Harper, Ralph
Novak, Barbara, 30
Nowak, Elke, 166–67

O'Connell, Nicholas, 149
O'Grady, John, 46, 192n19
Oates, Joyce Carol, 158–59
Orr, David, 130, 148
Otis, Elizabeth, 194n12
Overland Monthly, 36

Papin, Liliane, 74, 156
patch dynamics, 13, 152–53, 156, 161, 188n8. *See also* Loucks, O. L.; metastability; Wu, J.
Paul, Sherman, 154, 189n9
Peattie, Donald Cultoss, 60, 193n11
Peck, H. Daniel, 39
Pelly, David F., 166
Pennsylvania College for Women (Chatham College), 81
perception of relations, 10, 12, 17, 19, 27, 54, 62, 63, 69, 79, 86, 98, 100, 104, 105, 113, 137, 155, 170, 185
Perez, Betty L., 60
Peterson, Anna L., 17, 92, 197n22
Philippon, Daniel J., 190n3, 192n20; *Conserving Words: How American Nature Writers Shaped the Environmental Movement*, 7, 9, 24, 26, 30
Phillips, Dana, 41, 196n6; *The Truth of Ecology: Nature, Culture, and Literature in America*, 7–8, 9, 188n7, 202n1
Pinchot, Gifford, 27
Plumwood, Val, 195n2
Poirier, Richard, 120, 165, 200n5, 202n7

pragmatist ecology, 1–19, 34–36, 44, 51, 56, 59, 63, 64–66, 68, 74, 79, 80, 81, 86, 88, 89, 93, 96, 108, 115, 124, 125, 132, 145, 146–47, 149, 152, 153, 157, 158, 159, 162, 165, 167, 168, 170, 171, 174, 175, 177, 182–83, 184–85

Railsback, Brian E., 53, 193n7
Reuckert, William, 152
Ribicoff, Abraham, 195n3
Ricketts, Edward, 7, 9, 51–77, 79, 88, 92, 93, 183, 193n11; *Between Pacific Tides*, 183, 197n16; *Sea of Cortez*, 54–77, 79, 88
Ritter, William E., 53
Robinson, David, 189n9
Rogers, Steven B.: *A Gradual Twilight: An Appreciation of John Haines*, 200n2
Roorda, Randall, 44, 57, 119, 131, 132, 135, 147, 203n3
Rosenthal, Sandra B., 194n14
Ross-Bryant, Lynn, 203n5
Runte, Alfred, 190n5
Ruskin, John, 40, 191n17

St. Armand, Barton Levi, 28, 118
Saturday Review of Literature, 79
Scheese, Don, 191n9
Science, 56
self/identity, 28–32, 33–35, 41, 43, 102, 115, 131–32, 137, 144, 164, 191n10
Sevareid, Eric, 195n3
Sheffield, Carlton, 61, 194n13
Shelley, Percy Bysshe, 187n2, 202n3
Shusterman, Richard, 36, 190n8
Sierra Club, 22, 27
Silko, Leslie Marmon, 198n27
Siporin, Ona, 183
Slovic, Scott, 84, 144, 161, 162, 187n4
Smuts, Jan C., 53, 61
Snyder, Gary, 8–9, 174
Solnit, Rebecca, 29–30, 191n9
Stein, Roger B., 40, 41
Steinbeck, Carol, 55
Steinbeck, Elaine, 192n5

Steinbeck, John, 7, 9, 51–77, 79, 88, 92, 93, 145, 183, 193n11; *Cup of Gold*, 51; *East of Eden*, 52, 77, 192n4; *The Grapes of Wrath*, 51, 52, 55, 61, 68, 200n6; *The Log from the Sea of Cortez*, 55; *The Pastures of Heaven*, 51; *Sea of Cortez*, 54–77, 79, 88; *Steinbeck: A Life in Letters*, 192n5; *To a God Unknown*, 51, 192n1, 204n7
Steinbeck, Mary, 53
Stewart, Frank, 41, 195n3
Sweet, Timothy, 112

Tallmadge, John, 179, 202n6
Tanner, Tony, 158
Tatum, Stephen, 75, 154–55
Taylor, Bob Pepperman, 197n24
Taylor, C. V., 53
Thoreau, Henry David, 17, 27, 30, 33, 36, 39–40, 107, 191n13; *Journal*, 40; "Natural History of Massachusetts," 91; *Walden*, 14–16, 45, 131, 194n18; "Walking," 36–37, 191n13
Timmerman, John H., 192n2
Trueblood, Valerie, 199n2
Turner, Frederick, 48, 49, 190n1
Turner, Jack, 66

Uncle Tom's Cabin (Harriet Beecher Stowe), 80, 195n3
Up from Slavery (Booker T. Washington), 80
Usher, Peter, 166

Van Wyck, Peter C., 146
Ventor, J. Craig, 193n9
Virgil, 112
Vlessides, Mike, 166
Voyageur, Cora, 167

Wallace, David Raines: "What Is to Be Done with the Biosphere," 68
Walls, Laura Dassow, 191n16
Wesling, Donald, 41
West, Michael, 189n9
Westbrook, Robert B.: *John Dewey and American Democracy*, 188n5; *Democratic Hope: Pragmatism and the Politics of Truth*, 188n5
Weston, Anthony, 100, 106; "Beyond Intrinsic Value: Pragmatism in Environmental Ethics," 96–97
White, Fred D., 83–84
White, Lynn, Jr., 201n2
White, Richard, 140
Whitman, Walt, 51
Wild, Peter, 199n1
Wilkins, Thurman, 190n1
Williams, Raymond, 126–27
Williams, Terry Tempest, 7, 67–68, 170–85; John Dewey in, 19; *Desert Quartet: An Erotic Landscape*, 204n7; "A Eulogy for Edward Abbey," 183; *The Open Space of Democracy*, 6, 19, 176, 179, 185, 188n6; *Refuge: An Unnatural History of Family and Place*, 92, 172–85; "The Spirit of Rachel Carson," 183, 197n19; *An Unspoken Hunger*, 204n7; "Yellowstone: The Erotics of Place," 203n7
Williams, William Carlos, 112
Wilson, Edmund, 55
Wilson, Vanez: *Bear River: A National Wildlife Refuge*, 178–79
Wilson, Woodrow, 25
Wolfe, Linnie Marsh, 190n1; *John of the Mountains: The Unpublished Journals of John Muir*, 26, 30, 37, 39, 48, 49, 82
Worster, Donald, 27, 190n5; *Nature's Economy*, 81
Wu, J., 188n8; "From Balance of Nature to Hierarchical Patch Dynamics: A Paradigm shift in Ecology," 152–53. *See also* Loucks, O. L.; metastability; patch dynamics
Wynn, Graeme, 161

Yosemite National Park, 24
Yosemite Valley, 24, 25, 27, 37, 42

Zuelke, Karl, 176